Intervening Sequences
in Evolution and Development

Intervening Sequences in Evolution and Development

Edited by

EDWIN M. STONE
THE UNIVERSITY OF IOWA
COLLEGE OF MEDICINE

ROBERT J. SCHWARTZ
BAYLOR COLLEGE OF MEDICINE

New York   Oxford
OXFORD UNIVERSITY PRESS
1990

Oxford University Press

Oxford New York Toronto
Delhi Bombay Calcutta Madras Karachi
Petaling Jaya Singapore Hong Kong Tokyo
Nairobi Dar es Salaam Cape Town
Melbourne Auckland

and associated companies in
Berlin Ibadan

Copyright © 1990 by Oxford University Press, Inc.

Published by Oxford University Press, Inc.,
200 Madison Avenue, New York, New York 10016

Oxford is a registered trademark of Oxford University Press

All rights reserved. No part of this publication may be reproduced,
stored in a retrieval system, or transmitted, in any form or by any means,
electronic, mechanical, photocopying, recording, or otherwise,
without the prior permission of Oxford University Press.

Library of Congress Cataloging-in-Publication Data
Intervening sequences in evolution and development
edited by Edwin M. Stone and Robert J. Schwartz.
p. cm. Includes bibliographies and index. ISBN 0-19-504337-5
1. Chemical evolution. 2. Molecular genetics.
I. Stone, Edwin M.
II. Schwartz, Robert J. (Robert Joel)
[DNLM: 1. Evolution. 2. Genetics, Biochemical.
3. Molecular Biology. 4. Proteins—physiology.
QH 430 I603] QH325.I54 1989 574.87'328—dc20
DNLM/DLC for Library of Congress
89-8822 CIP

9 8 7 6 5 4 3 2 1

Printed in the United States of America
on acid-free paper

Preface

In the mid-1970's, most molecular biologists felt quite certain that the genes of eukaryotes would prove to be organized and regulated in ways that were very similar to those of prokaryotes. The precise linear relationship between bacterial genes and their proteins seemed so reasonable that there was little doubt it would be universal. Similarly, the regulatory elegance of bacterial operons led most investigators to expect that eukaryotic genes would be regulated in similar ways. These expectations were bolstered by a prevailing concept of 1970s evolutionary biology: that eukaryotes arose from an endosymbiotic fusion of prokaryote-like primordial organisms. No one dreamed that the "simple prokaryote" that gave rise to the first eukaryotic nucleus contained genes whose coding regions were divided by stretches of noncoding sequence that had to be perfectly excised at the RNA level before the genes could be expressed. It was with considerable skepticism that investigators looked at genomic Southern blots that seemed to indicate that eukaryotic genes were longer than their corresponding mRNAs and that these extra sequences were interspersed within the coding sequences. By 1978, however, the existence of these "intervening sequences" and the requisite RNA splicing apparatus had been firmly established in several eukaryotic systems.

The potential evolutionary utility of this "genes-in-pieces" genomic organization was quickly perceived by Walter Gilbert.[1] He recognized that shuffling of functional gene segments among various parent genes would be potentiated if these meaningful segments (exons) were separated by unimportant DNA sequences (introns) in which unequal recombination events could occur without altering the phase or the length of the coding segments. Although Gilbert did not explicitly restrict his ideas about mosaic gene structure to eukaryotes, most people who read his article in 1978 did so with the preconception that introns and RNA splicing were complexities that had been *added* to eukaryotes following their divergence from primoridal prokaryotes. Ford Doolittle[2] was quick to point out how difficult it was to envision any evolutionary advantage in dividing previously intact genes. He reminded us that no matter how beneficial mosaic gene organization might be once it existed, it was non-Darwinian to sug-

---

[1] Gilbert, W. (1978). *Nature* 271: 501.
[2] Doolittle, W. F. (1978). *Nature* 272: 581.

gest that organisms would tolerate the immediate disadvantage of split genes for some later benefit. Instead, he proposed that the primordial organism that preceded modern eukaryotes and prokaryotes (the progenote of Woese and Fox) had RNA splicing machinery and that introns participated in the evolution of the earliest genes.

Colin Blake,[3] also in 1978, expanded Gilbert's ideas into the realm of protein structure. If introns frequently separate fully functional gene segments that could have been shared among genes during evolution, Blake suggested that such introns should divide the coding sequence at points corresponding to domain borders in the resulting proteins. This idea provided a mechanism for testing the hypothesis that introns facilitated the evolution of some genes. In 1983, Blake[4] pointed out that Doolittle's idea of progenotic splicing could be addressed by studying the relationship of introns and domain borders in ancient genes (present before the divergence of eukaryotes and modern prokaryotes). He wrote: "the critical point is, of course, whether early cells had split genes; and this question might be finally settled if exon-dependent evolution can be established for eukaryotic proteins which have defined homologues in prokaryotes."

When Blake's note appeared in 1983, our laboratory was studying the developmental expression of various genes in avian muscle. We had cloned the glyceraldehyde-3-phosphate dehydrogenase (GAPDH) gene from the chicken and sequenced it to compare its 5' regulatory sequences with those of various actin genes. Blake's last sentence quoted above made us realize that we had a gene sequence that was relevant to the hypothesis of progenotic RNA splicing. GAPDH is a very ancient protein and has been highly conserved from bacteria to man. Moreover, it was the subject for much early X-ray crystallography and thus the domain borders necessary for a Blake-type analysis were known. We compared the intron locations in the chicken GAPDH gene to the domain borders worked out by Rossmann and his co-workers years earlier, and found a relationship that we felt argued strongly for the presence of mosaic genes in primordial organisms.

In recent years, much has been learned about the chemistry of RNA splicing in various systems, which has added weight to the idea that this process could have existed in the progenote. In addition, a nonevolutionary function for introns and RNA-splicing has been found to be widespread throughout the biological kingdoms: the generation of multiple protein isoforms from a single gene via alternative splicing of pre-mRNAs.

As we began to organize the task of writing a book on the origin and function of introns, our objective was to provide several different perspectives on the topic by eliciting chapters from scientists with varied backgrounds including: protein structure, evolutionary biology, eukaryotic gene regulation, RNA chemistry, and developmental biology. Our specific purposes in writing this book are to: (1) fully explore several types of evidence that suggest that intervening sequences and RNA splicing have played an important role in the con-

---

[3] Blake, C. C. F. (1978). *Nature* 273: 267.
[4] Blake, C. C. F. (1983). *Nature* 306: 535.

struction and refinement of genes throughout evolution; (2) present an up-to-date review of the known mechanisms of RNA splicing; and (3) discuss the participation of introns in the control of gene expression.

In the first chapter of the book, Rossmann reviews the studies of dehydrogenases that led to the realization (years before introns were discovered) that protein domains were shuffled among related enzymes during evolution. In Chapter 2, Holland and Blake develop the idea that gene structure should be mirrored in protein structure, and present a comprehensive review of the known examples of proteins that seem to have evolved with the aid of intron-mediated domain shuffling. In the third chapter, Doolittle addresses the origin of intervening sequences and presents his idea that intervening sequences and RNA splicing were probably present in the progenote. In addition, he critically examines some of the "common sense" ideas of the evolutionary utility of introns against a rigorous background of classic evolutionary biology. In Chapter 4, we present evidence from the study of ancient dehydrogenase genes to support Doolittle's idea of RNA splicing in primordial organisms. We also discuss the interpretation of alpha carbon distance plots in some detail and utilize these plots to demonstrate graphically the concordance of introns and domain borders in the dehydrogenases. Lastly, we synthesize the ideas of a number of other investigators to propose a mechanism, dependent upon RNA splicing, whereby functional enzymes could have evolved *before* their genes. In the fifth chapter, Stein, Scott, and O'Malley present a historical perspective of the discovery of introns and discuss the early hypotheses about their function. They also summarize their evidence for intron-mediated duplication of domains during the evolution of ovomucoid. In the sixth chapter, Perlman, Peebles, and Daniels define and discuss all four types of introns, including those in chloroplast and mitochondrial genes. They present an extensive review of the current information about the splicing mechanisms used to excise each type, as well as the evolutionary relationships among them. In Chapter 7, Smith, Knaack, and Nadal-Ginard review the phylogenetically widespread use of alternative intron splicing as a means of generating multiple protein isoforms from a single gene. They also review a number of examples in which alternative splicing is used to regulate the function of genes during organismal development.

We feel that this book will be of interest to readers ranging from graduate students to senior scientists who wish to broaden their understanding of intervening sequences and the important roles they play in evolution and development.

We would like to thank our contributing authors for the enormous amount of effort they spent writing their chapters. We would also like to thank William Curtis of Oxford University Press for encouraging us to undertake this project, as well as his colleague, Stanley George for his tireless effort in bringing it to completion.

Iowa City, Iowa   E. M. S.
Houston, Texas   R. J. S.

Contents

Contributors, xi

1. Introductory Comments on the Function of Domains in Protein Structure, 3
   MICHAEL G. ROSSMANN

2. Proteins, Exons, and Molecular Evolution, 10
   S. K. HOLLAND AND C. C. F. BLAKE

3. Understanding Introns: Origins and Functions, 43
   W. FORD DOOLITTLE

4. Intron-Dependent Evolution of Progenotic Enzymes, 63
   EDWIN M. STONE AND ROBERT J. SCHWARTZ

5. Intervening Sequences in Molecular Evolution, 92
   JOSEPH P. STEIN, MAXWELL J. SCOTT, AND BERT W. O'MALLEY

6. Different Types of Introns and Splicing Mechanisms, 112
   PHILIP S. PERLMAN, CRAIG PEEBLES, AND CHARLES DANIELS

7. Alternative mRNA Splicing in the Generation of Protein Diversity and the Control of Gene Expression, 162
   CHRISTOPHER W. J. SMITH, DAVID KNAACK, AND BERNARDO NADAL-GINARD

Index, 197

Contributors

C. C. F. Blake
Laboratory of Molecular Biophysics,
The Rex Richards Building,
Department of Zoology,
University of Oxford,
South Parks Road, Oxford
OX1 3QU, England.

Charles Daniels
Department of Microbiology,
The Ohio State University,
Columbus, Ohio 43210.

W. Ford Doolittle
Department of Biochemistry,
Dalhousie University,
Halifax, Nova Scotia
B3H 4H2, Canada.

S. K. Holland
Laboratory of Molecular Biophysics,
The Rex Richards Building,
Department of Zoology,
University of Oxford,
South Parks Road, Oxford
OX1 3QU, England.

David Knaack
Laboratory of Molecular
  and Cellular Cardiology,
Howard Hughes Medical Institute,
Department of Cardiology,
  Children's Hospital, and
Departments of Pediatrics,
Physiology and Biophysics,
Harvard Medical School,
Boston, Massachusetts 02115.

Bernardo Nadal-Ginard
Laboratory of Molecular
  and Cellular Cardiology,
Howard Hughes Medical Institute,
Department of Cardiology,
  Children's Hospital, and
Departments of Pediatrics,
Physiology and Biophysics,
Harvard Medical School,
Boston, Massachusetts 02115.

Bert W. O'Malley
Department of Cell Biology,
Baylor College of Medicine,
One Baylor Plaza,
Houston, Texas 77030.

Craig Peebles
Department of Biological Sciences,
University of Pittsburgh,
Pittsburgh, Pennsylvania 15260.

Philip S. Perlman
Department of Molecular Genetics,
The Ohio State University,
Columbus, Ohio 43210.

Michael G. Rossmann
Department of Biological Sciences,
Purdue University,
West Lafayette, Indiana 47907.

Robert J. Schwartz
Department of Cell Biology,
Baylor College of Medicine,
One Baylor Plaza,
Houston, Texas 77030.

Maxwell J. Scott
Department of Biology,
University of North Carolina,
Coker Hall,
Chapel Hill, North Carolina 27514.

Christopher W. J. Smith
Laboratory of Molecular
 and Cellular Cardiology,
Howard Hughes Medical Institute,
Department of Cardiology,
 Children's Hospital, and
Departments of Pediatrics,
Physiology and Biophysics,
Harvard Medical School,
Boston, Massachusetts.

Joseph P. Stein
Division of Endocrinology/Room 4200,
University of Texas Medical School,
6431 Fannin Street,
Houston, Texas 77225.

Edwin M. Stone
Molecular Ophthalmology Laboratory,
Department of Ophthalmology,
The University of Iowa
College of Medicine,
Iowa City, Iowa 52242.

Intervening Sequences
in Evolution and Development

1

Introductory Comments on the Function of Domains in Protein Structure

MICHAEL G. ROSSMANN

It is difficult to return to the concepts prevalent only a few years ago, let alone three decades ago. Accumulated knowledge becomes so much a part of everyday life that a view of the scientific horizons in the mental frame of yesterday is almost impossible. The ignorance prevalent only a few years ago is often so inconceivable that it is frequently denied even by those who participated in the accumulation of knowledge and whose work contributed to a more fundamental understanding. As I write this introduction I find it astonishing how much has been accomplished in the last 30 years.

I remember distinctly how, in 1958 when I joined the MRC Unit of Molecular Biology in Cambridge, Max Perutz felt it necessary to convince me that hemoglobin is homogenous and has a unique structure, although Sanger had already sequenced insulin and Kendrew had determined a crude 6-Å structure of myoglobin. I remember the immediate satisfaction that accompanied the demonstration that seal myoglobin had essentially the same three-dimensional structure as sperm whale myoglobin (Scouloudi, 1959) long before there was any knowledge of amino acid sequences, for there had been much discussion of the dramatic effect that even a single changed residue might have on conformation. For instance, Ingram had shown that there was only one amino acid change between normal and sickle-cell hemoglobin (Ingram, 1956). Given that view, it was hardly conceivable that the α and β chains of hemoglobin and the myoglobin molecule could be related. The discovery to the contrary (Perutz et al., 1960) had, therefore, a profound effect on the conceptual framework of molecular biology. Here was the idea that proteins of similar function may have had a common precursor and that evidence of this evolutionary process resided in structure. It was only later (Edmundson, 1965; Braunitzer et al., 1964) that it was demonstrated that there was also homology in the amino acid sequences of myoglobin and the hemoglobin polypeptide chains. With these concepts in place, it was possible for Fitch and Margoliash (1967) to use the cytochrome c molecule as a tool to derive a phylogeny of eukaryotic organisms in excellent agreement with fossil records. There then gradually emerged the concept that structure, required for the control of function, is conserved over periods of time during which the primary sequence is changing much faster. Indeed, there is frequently little evidence of a common primordial amino acid sequence, while structure and function are little changed. That leaves us with the still largely unanswered question, "What properties of the amino acids are essential in the sequence to maintain a given fold?"

In spite of the profound effect the hemoglobin structure had on the concepts of protein conformation and evolution, it did not seem at all obvious that, for instance, the isozymes of lactate dehydrogenase (LDH) (Markert and Appella, 1963) might be homologous and have related structure, far less that different NAD-dependent dehydrogenases should have a common evolutionary origin. Nevertheless, it was these possibilities that in part led me (Rossmann et al., 1967) to study LDH and glyceraldehyde-3-phosphate dehydrogenase (GAPDH). Certainly Carl Branden must have had the same anticipations in his work on liver alcohol dehydrogenase (LADH). In a 1969 meeting on pyridine nucleotide-dependent enzymes, held in Konstanz, Carl and I compared our low-resolution structures of ADH and LDH (Branden and Rossmann, 1970). Our optimistic interpretation suggested a structural similarity, the full significance of which was to become apparent only a few years later, when the atomic resolution structures of cytoplasmic malate dehydrogenase (Tsernoglou et al., 1972), GAPDH (Buehner et al., 1973), LDH (Adams et al., 1970), and LADH (Branden et al., 1973) were all available. It then became apparent that there was a common "NAD binding domain" and a mostly quite different "catalytic domain," the latter providing residues important for binding substrate and for catalysis. The similarity in which NAD bound to its domain, the conservation of functionally important residues (e.g., an aspartate residue always bound the $2'$ oxygen of the adenine ribose), and the conservation of residues thought to be important in giving folding signals (e.g., a glycine at the beginning of the "αB" helix) were quickly explored (Rossmann et al., 1974; Ohlsson et al., 1974).

The domain concept was not entirely new to protein structure. It was introduced in the analysis of immunoglobulins (Hill et al., 1966) with its partially repeating sequences within the heavy and light chains. However, even before the comparison of LDH with other dehydrogenases was possible, it was obvious (Adams et al., 1970) that there were quite different architectural principles involved in folding the first and second parts of the LDH monomer. The first "domain" was based on a parallel β-pleated sheet with returning parallel α-helices, whereas the second domain was constructed from antiparallel β-pleated ribbons. The two domains, although in contact, did have some spatial separation. Furthermore, the protein architecture could be related to function, a concept that can readily be related to everyday experiences. This analogy appealed to me strongly in light of the indoctrination I had received at an early age from my mother, who had been a Bauhaus student, and from an uncle who had been a renowned classical Greek scholar.

These ideas were still evolving (Liljas and Rossmann, 1974; Rossmann et al., 1975) when a number of structures were determined, all of which had an architecture similar to the NAD binding domain in dehydrogenases. Most of these structures bound mononucleotides in much the same way as AMP or nicotinamide mononucleotide bound to LDH, suggesting a primordial mononucleotide binding fold. These structures had three parallel β-strands with returning α-helices, and their fold was always of the same hand (Rao and Rossmann, 1973). It was not until later that Hol suggested a physical reason why these structures bound nucleotides in terms of the dipole moments of the helices (Hol et al., 1978). The prevalence of the nucleotide structures created much interest and, in particular, Sternberg and Thornton (1976) and Richardson (1977) analyzed $\beta-\alpha-\beta$ folds.

Here, then, was a vast extension of the architectural principles relating fold and function. Furthermore, these generalized nucleotide binding structures exhibited similar amino acid patterns ("fingerprints") that permitted the prediction that, for instance, tRNA synthetases (which bind ADP) would have similar architecture—a prediction that turned out to be right (Irwin et al., 1976; Bhat et al., 1982; Risler et al., 1981).

But sweeping new concepts seldom go without challenge, as is indeed necessary for the orderly development and the separation of fact from hypothesis. How was it possible to establish that some structures were similar to each other in the light of numerous insertions and deletions? How complex must a structure be to suggest a common primordial origin rather than a preferred folding pattern such as an α-helix or even an $\alpha-\beta-\alpha$ fold? Is it possible to recognize unequivocably a domain within a monomer on the basis of spatial separation (Wodak and Janin, 1980) or by comparison with other structures? It is inconceivable today to realize that, for a while, there were those who even doubted the similarity of the NAD binding domains in dehydrogenases. There were many more who could not see that flavodoxin is more similar to LDH than hemoglobin is to LDH. For this reason I developed quantitative methods for comparing structures (Rao and Rossmann, 1973; Rossmann and Argos, 1975, 1976) that not only established the degree of similarity between any two structures but also searched for similarity where none was previously known. These techniques were used also to find structurally and functionally similar domains, such as the heme binding domain in globins, cytochrome c's and cytochrome b's; the polysaccharide binding domains in various lysozymes (Rossmann and Argos, 1976); or the cell attachment domains in lectins and viruses (Argos et al., 1980).

The work on lysozymes encouraged Remington and Matthews (1978) to develop their own techniques, which, in conjunction with the methods developed at Purdue University, now provide generally accepted standards of comparison (Matthews and Rossmann, 1985). Table 1.1, for instance, provides a series of

Table 1.1. Structural Comparisons

Proteins*		Number of Residues		Number of Equivalences	Percentages of Equivalences		MBC/C†
Mol. 1	Mol. 2	Mol. 1	Mol. 2		Mol. 1	Mol. 2	
Hb(β)	Hb(α)	146	141	139	95	99	—
SBMV (C)	TBSV (C)	221	198	179	81	90	—
V_H	V_L	110	110	77	70	70	—
GAPDH(NAD)	LDH(NAD)	148	144	96	65	67	1.24
Hb(β)	Cyt. b_5	146	86	58	40	67	1.29
Con A	TBSV (P)	237	110	68	29	62	—
T4L	HEWL	164	129	78	48	60	1.53
CAP(DNA)	Cro	73	66	31	42	47	—
LDH(NAD)	Flavod.	144	138	39	27	28	1.23

*Hb(α) and Hb(β), α and β chains of horse hemoglobin; SBMV (C), C subunit of southern bean mosaic virus; TBSV; (C), C subunit of S domain of tomato bushy stunt virus; V_H and V_L, variable domains of heavy and light immunoglobulin chains; Con A, concanavalin A: GAPDH(NAD), nicotinamide adenine dinucleotide binding domain of glyceraldehyde-3-phosphate dehydrogenase; LDH(NAD), NAD binding domain of lactate dehydrogenase; Cyt. b_5, cytochrome b_5; T4L, bacteriophage T4 lysozyme; HEWL, hen egg-white lysozyme; CAP(DNA), DNA binding domain of catabolite gene activator protein; Cro, *cro* repressor protein; Flavod., flavodoxin.
†Minimum base change per codon.

benchmarks relative to which other comparisons can be calculated. Perhaps the single most useful parameter is the percentage of residues that can be spatially equivalenced (for detailed definitions see Rossmann and Argos, 1977). However, other parameters—such as similarity of substrate or prosthetic group binding sites, similarly positioned catalytic residues, complexity of the structure, and homology of equivalenced amino acids—are all important. At one end of Table 1.1 is the equivalence between the α and β chains of hemoglobin, which have numerous characteristics, to make divergent evolution from a common ancestor highly probable. At the other end of the scale are the lysozyme comparison (low percentage of equivalent residues) and the DNA binding proteins (a simple structure), either of which could just possibly be the result of convergent evolution to a functionally useful structure.

With this experience it is possible to attempt to define a "domain." However, rather than giving a simple definition this is best done by describing a set of domain properties, not all of which need hold in a given case. The following are the most important properties:

1. A domain is structurally compact, of a single type of folding architecture (e.g., a parallel β-pleated sheet), and spatially distinct from other domains in the polypeptide.
2. There is structural similarity and possibly sequence homology to other domains, either in the same polypeptide chain or in another molecule.
3. There is similarity in function to domains of similar structure. Such similarity can be expressed as:
   a. Similar positioning of residues important for catalysis.
   b. Similar positioning of substrate, coenzyme, or prosthetic groups.

When a protein domain is defined in this manner, it follows that the reactive center of enzymes should be situated between domains, with each domain providing a simple component of the functions required for the more complex enzyme (Liljas and Rossmann, 1974; Rossmann and Argos, 1981). Here, then, is also evidence for gene fusion, which provides the possibility of more rapid evolution of proteins or enzymes by the exchange of whole DNA components, each coding for a complete domain. For instance, a difference in dehydrogenases could be produced by fusing the gene of an NAD binding domain to any number of genes representing different catalytic domains (Rossmann et al., 1972, 1974). Although the dehydrogenases provided the first strong evidence for gene shuffling of domains, numerous other examples are known. Inspection of a new structure is now invariably in terms of recognizing the architectural character of the various domains and perhaps relating them to previously known structures of similar function. For instance, the final domain of fungal catalase is closely similar to flavodoxin, although no FMN binding capacity has yet been associated with this domain (Vainshtein et al., 1981; Melik-Adamyan et al., 1986).

Quite independently and somewhat later it was discovered that eukaryotic genes have lengthy intervening unexpressed sequences (introns) of lengths seldom less than about 180 base pairs dividing the coding sequence of the gene into pieces (exons). Gilbert (1978) and Artymiuk et al. (1981) then asked whether exons were the genetic basis of protein domains. The necessary gene splicing could be an ideal mechanism for the gene shuffling and fusion implied by the observation of domain structure. Early results appeared to favor this hypothesis. For instance, the central globin gene of vertebrates related to the central heme binding domain (Rossmann and Argos, 1975), and the middle two exons of lysozymes related to the polysaccharide binding domain (Artymiuk et al., 1981). Later results, however, showed that the situation was far more complex, requiring additional ad hoc explanations (Craik et al., 1982). Up-to-date analyses of the current information of the relation between exons and protein domains are given elsewhere in this book.

The frequent occurrence of the mononucleotide binding domain structure produced some vigorous controversy in the 1970s. Nevertheless, the results and concepts that were thus generated, and that are generally applicable to all protein structure, are hardly in dispute today. I am, therefore, fascinated by the effect that the recent burst of virus structures will have on our structural concepts. Essentially all spherical viruses analyzed to high resolution contain the same coat protein structural domain (Rossmann et al., 1985; Hogle et al., 1985). Argos et al. (1980) suggested that primordial domains with the ability to bind to cell surface structures have been utilized by viruses to recognize specific host tissue. Studies of cell receptors have been difficult both because they are membrane bound and because they are present only in small quantities. Yet their biological function is as universally important as nucleotide binding is to many enzymes. Indeed in writing this final paragraph I am treading on ground as controversial as the recognition of similar proteins and their possible common origin was a dozen years ago.

REFERENCES

Adams, M. J., Ford, G. C., Koekoek, R., Lentz, P. J. Jr., McPherson, A., Jr., Rossmann, M. G., Smiley, I. E., Schevitz, R. W., and Wonacott, A. J. (1970). Nature (London) 227:1098–1103.
Argos, P., Tsukihara, T., and Rossmann, M. G. (1980). J. Molec. Evol. 15:169–179.
Artymiuk, P. J., Blake, C. C. F., and Sippel, A. E. (1981). Nature (London) 290:287–288.
Bhat, T. N., Blow, D. M., Brick, P., and Nyborg, J. (1982). J. Molec. Biol. 158:699–709.
Branden, C. I., and Rossmann, M. G. (1970). In Pyridine Nucleotide Dependent Dehydrogenases, Sund, H., ed. Springer-Verlag, Berlin, pp. 133–134.
Branden, C. I., Eklund, H., Nordstrom, B., Boiwe, T., Soderlund, G., Zeppezauer, E., Ohlsson, I. and Akeson, A. (1973). Proc. Natl. Acad. Sci. USA 70:2439–2442.
Braunitzer, G., Hilse, K., Rudloff, V., and Hilchman, N. (1964). Adv. Prot. Chem. 19:87–95.
Buehner, M., Ford, G. C., Moras, D., Olsen, K. W., and Rossmann, M. G. (1973). Proc. Natl. Acad. Sci. USA 70:3052–3054.
Craik, C. S., Sprang, S., Fletterick, R., and Rutter, W. J. (1982). Nature (London) 299:180–182.
Edmundson, A. B. (1965). Nature (London) 205:883–887.
Fitch, W. M., and Margoliash, E. (1967). Science 155:279–284.
Gilbert, W. (1978). Nature (London) 271:501.
Hill, R. L., Delaney, R., Fellows, R. E., Jr., and Lebovitz, H. E. (1966). Proc. Natl. Acad. Sci. USA 56:1762–1769.
Hogle, J. M., Chow, M., and Filman, D. J. (1985). Science 229:1358–1365.
Hol, W. G. J., van Duijnen, P. T., and Berendsen, H. J. C. (1978). Nature (London) 273:443–446.
Ingram, V. M. (1956). Nature (London) 178:792.
Irwin, M. J., Nyborg, J., Reid, B. R., and Blow, D. M. (1976). J. Molec. Biol. 105:577–586.
Liljas, A., and Rossmann, M. G. (1974). J. Molec. Biol. 85:177–181.
Markert, C. L., and Appella, E. (1963). Ann. N.Y. Acad. Sci. 103:915–929.
Matthews, B. W., and Rossmann, M. G. (1985). Meth. Enzymol. 115:397–420.
Melik-Adamyan, W. R., Barynin, V. V., Vagin, A. A., Borisov, V. V., Vainshtein, B. K., Fita, I., Murthy, M. R. N., and Rossmann, M. G. (1986). J. Molec. Biol. 188:63–72.
Ohlsson, I., Nordstrom, B., and Branden, C. I. (1974). J. Molec. Biol. 89:339–354.
Perutz, M. F., Rossmann, M. G., Cullis, A. F., Muirhead, H., Will, G., and North, A. C. T. (1960). Nature (London) 185:416–422.
Rao, S. T., and Rossmann, M. G. (1973). J. Molec. Biol. 76:241–256.
Remington, S. J., and Matthews, B. W. (1978). Proc. Natl. Acad. Sci. USA 75:2180–2184.
Richardson, J. S. (1977). Nature (London) 268:495–500.
Risler, J. L., Zelwer, C., and Brunie, S. (1981). Nature (London) 292:384–386.
Rossmann, M. G., and Argos, P. (1975). J. Biol. Chem. 250:7525–7532.
Rossmann, M. G., and Argos, P. (1976). J. Molec. Biol. 105:75–95.
Rossmann, M. G., and Argos, P. (1977). J. Molec. Biol. 109:99–129.
Rossmann, M. G., and Argos, P. (1981). Ann. Rev. Biochem. 50:497–532.

Rossmann, M. G., Jeffery, B. A., Main, P., and Warren, S. (1967). Proc. Natl. Acad. Sci. USA 57:515–524.
Rossmann, M. G., Ford, G. C., Watson, H. C., and Banaszak, L. J. (1972). J. Molec. Biol. 64:237–249.
Rossmann, M. G., Moras, D., and Olsen, K. W. (1974). Nature (London) 250:194–199.
Rossmann, M. G., Liljas, A., Branden, C. I., and Banaszak, L. J. (1975). In The Enzymes. Vol: XI, 3rd ed. Boyer, P. D., ed., Academic Press, New York, pp. 61–102.
Rossmann, M. G., Arnold, E., Erickson, J. W., Frankenberger, E. A., Griffith, J. P., Hecht, H. J., Johnson, J. E., Kamer, G., Luo, M., Mosser, A. G., Rueckert, R. R., Sherry, B., and Vriend, G. (1985). Nature (London) 317:145–153.
Scouloudi, H. (1959). Nature (London) 183:374–375.
Sternberg, M. J. E., and Thornton, J. M. (1976). J. Molec. Biol. 105:367–382.
Tsernoglou, D., Hill, E., and Banaszak, L. J. (1972). Cold Spring Harbor Symp. Quant. Biol. 36:171–178.
Vainshtein, B. K., Melik-Adamyan, W. R., Barynin, V. V., Vagin, A. A., and Grebenko, A. I. (1981). Nature (London) 293:411–412.
Wodak, S. J., and Janin, J. (1980). Proc. Natl. Acad. Sci. USA 77:1736–1740.

2
Proteins, Exons, and Molecular Evolution

S. K. HOLLAND AND C. C. F. BLAKE

The discovery of the mosaic gene structure in eukaryotes has stimulated much research and discussion regarding its influence on cellular evolution and the implications for the proteins specified by structural genes. Observed correlations between protein structure, function, and the split gene structure have prompted many interpretations of the origin and role of the mosaic gene in the development of the wide range of proteins found in present-day cellular organisms.

In bacteria, a structural gene, consisting of a discrete, contiguous stretch of DNA encodes the information to specify the manufacture of a single protein in a direct correspondence between the nucleotide and amino acid sequences. Such colinearity was assumed to apply to higher organisms, until 1977; when it became apparent that for eukaryotes the gene is split into short, coding regions called exons and long, noncoding regions called introns (Breathnach et al., 1977; Breathnach and Chambon, 1981). Following transcription of the gene, the introns are excised and the exons are ligated to form the mature messenger RNA (see Fig. 2.1).

This review will examine the growing evidence for the different hypotheses proposed to explain the origin and function of the mosaic gene, its correlation with protein structure, and the implications for cellular evolution.

EXON SHUFFLING—AN EVOLUTIONARY ADVANTAGE?

At the molecular level, it is generally accepted that present-day proteins descended from a relatively small number of primordial proteins, by the mechanism of gene duplication by recombination. Duplication of a structural gene

PROTEINS, EXONS, AND MOLECULAR EVOLUTION

Eucaryotic Genes

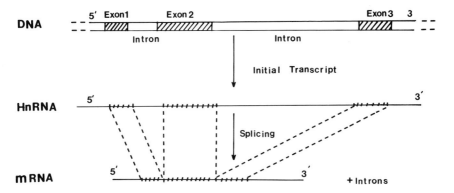

FIGURE 2.1. The structure of the eukaryotic gene and its transcription into RNA.

allows one copy to retain the original function while leaving the second copy free to mutate without the constraints imposed by natural selection. The occasional, advantageous, mutational event can then create either a modified version of the original protein or a protein with a new functional capability.

However, as structural examination of protein molecules extended into the range of larger, intracellular proteins, a feature was revealed that was difficult to fit into this simple evolutionary mechanism. Virtually all polypeptide chains with more than 200 amino acid residues and some with less were found to be structurally organized into two or more domains. Domains are regions of autonomous structure, with the appearance of a complete protein molecule, but linked by the polypeptide chain to one or more other domains. Furthermore, domains are usually found to be associated with a particular binding function, and where the protein in question is an enzyme, the active site is almost invariably located at the domain interface. These proteins give the impression of having evolved by the linking of two or more "proto-proteins," each carrying its own binding functions, by a mechanism involving gene fusion. In 1974, strong support was given to this idea by Rossmann's observation that the NAD-binding domains of four dehydrogenases have very similar structural organization and binding properties, whereas the second catalytic domains in these proteins were quite different from one another (Rossmann et al., 1974). Rossmann concluded that the NAD-binding domains were descended by duplication from a common ancestor and subsequently linked to different catalytic domains to form the individual dehydrogenases by gene fusion. A process of this kind, through its ability to combine different functions in the same protein molecule, would give a much greater resource in the production of novel protein molecules than duplication alone. What, however, is the general mechanism of gene fusion?

The answer, inconceivable in 1974, came from the realization during 1977 that eukaryotic genes were split into a number of coding regions embedded in a larger matrix on noncoding DNA. Gilbert (1978) was the first to realize that if the coding regions, or exons, encoded protein functions, then recombination within the large, noncoding regions, or introns, could reassort the functions in novel protein molecules. With continuous genes, such as those of present-day prokaryotes, the structural requirements of the proteins and the need to avoid frameshift mutations may restrict the target for successful gene recombination to very few bases. In contrast, the great length of introns in eukaryotic genes, often in the kilobase range, and the absence of a reading frame in these regions increase the target for recombination by up to three orders of magnitude (see Fig. 2.2). As Gilbert has pointed out, this size of target should ensure that intronic recombination is a significant factor in the evolution of the products of split genes.

The exon-shuffling hypothesis was extended by Blake, who suggested that exons encoded structural elements such as domains or smaller, supersecondary structures (Blake 1978, 1979). Proteins that fold into localized domains often contain similar patterns of substructures. Once a successful architecture has been adopted it is apparently used repeatedly. Although whole domains define function, they are subdivided into smaller supersecondary structures. Fusion of these units by linking small exons could produce the folded domain as efficiently as encoding it on one larger exon and also increase the permutations for producing new proteins, a process of significant evolutionary advantage. Such substructures could be adapted for different functions and transferred by the exon-shuffling mechanism as before, to create novel proteins composed of the different functional units present in preexisting proteins (see Fig. 2.3). An additional advantage of this role for exons is that as each unit displays folding

FIGURE 2.2. A diagram to show the targeting advantage for unequal crossover in a split gene.

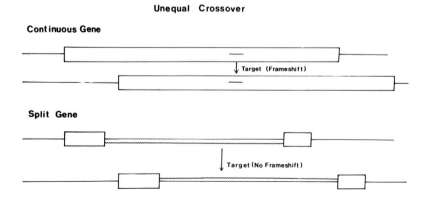

PROTEINS, EXONS, AND MOLECULAR EVOLUTION

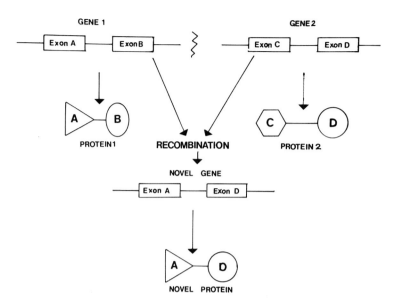

FIGURE 2.3. A schematic representation of the Gilbert–Blake exon-shuffling hypothesis.

autonomy, the need for development of new folding pathways in the novel proteins is eliminated and the probability of successful recombination is further enhanced. Notably the lack of integral structure in the exon-encoded fragment would reduce to almost zero the chances of a folded protein being created by linking such units together.

These theories offer a genetic mechanism to explain the high frequency of occurrence of multidomain proteins, domains that encode individual functions, and the repeated use of similar structural motifs in different proteins. The exon-shuffling hypothesis can also account for the rapidity of evolution from simple, primordial molecules to the complex and diverse range of proteins observed today. Comparison of gene structure and the three-dimensional structure of multifunctional proteins allows the hypothesis of exon shuffling to be tested directly. The observation of different genes encoding a single element of structure and function on the same exon in otherwise different proteins would prove the theory correct. As the number of three-dimensional protein structures is still small, most of the correlation of exons and protein structure is inferred from sequence homology.

EVIDENCE FOR EXON SHUFFLING, CORRELATION OF PROTEIN AND GENE STRUCTURES

The recent discovery of a group of eukaryotic proteins that display a complex distribution of similar functional domains indicates that their evolution pro-

14   INTERVENING SEQUENCES IN EVOLUTION AND DEVELOPMENT

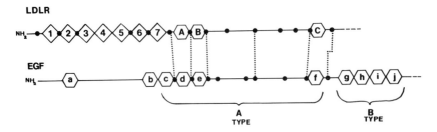

FIGURE 2.4. The gene structures of low-density lipoprotein receptor (LDLR) and epidermal growth factor (EGF). Diamonds represent domains homologous to Complement 9 domains and hexagons represent EGF units: A and B, as designated. Intron junctions are shown as dots, and identical splice junctions are shown interconnected by a dotted line.

ceeded by gene duplication and the free diffusion of domains into the proteins when required, rather than by descent from a common ancestor. The correspondence between protein domains and exon–intron boundaries in the genome in several proteins of the group appears to validate completely the exon-shuffling hypothesis.

The proteins include the blood coagulation and fibrinolytic serine proteases, a hormone precursor, a receptor, and an extracellular matrix protein. The domains are mostly small, about 40 amino acids long, with the homologies identifiable by the pattern of conserved disulfide bridges.

The epidermal growth factor precursor (EGFP) contains 10 repeats of a 40-residue, cysteine-rich sequence whose common patterns of disulfide bridges strongly suggest that they were derived from a common ancestor (see Fig. 2.4). (Gregory et al., 1977; Scott et al., 1983; Gray et al., 1983). Closer examination of the degree of sequence homology indicates that c, d, e, f units (designed EGF B units) are more closely related to each other than g, h, i, j units (designated EGF A units) (Doolittle et al., 1984). (The a and b units are more distantly related to all the other EGF units.) The patterns of disulfide bridges in the A- and B-type domains are similar, but not identical, indicating that they may be differentiated versions of the same structural form and that the B unit developed from the A unit.

Both types of EGF domain are found elsewhere in different combinations (see Fig. 2.5) (Doolittle et al., 1984). The EGF A unit is observed once in tissue plasminogen activator (TPA) (Pennica et al., 1983; Banyai et al., 1983); and urokinase (UK) (Steffens et al., 1982; Gunzler et al., 1982); and Complement C9 (DeScipio et al., 1984). It occurs twice in Factor XII (McMullen and Fujikawa, 1985), although at nonadjacent positions. It is also found in the vaccinia virus (Blomquist et al., 1984; Brown et al., 1985) and transforming growth factors (Marquardt et al., 1983; Derynck et al., 1984). In Factor IX (Katayama et al., 1979; Kurachi and Davie, 1982), Factor X (Enfield et al., 1980; Leytus et al., 1984; Fung et al., 1984), Protein C (Fernlund and Stenflo,

1982; Foster and Davie, 1984), and Protein Z (Hojrup et al., 1985), one copy of each type of EGF unit is found, in adjacent positions. Recently, the primary structure of Protein S has revealed that the protein contains a series of four adjacent EGF A units in the noncatalytic segment (Dahlbeck et al., 1986).

Of the gene structures known, Factor IX (Anson et al., 1984), TPA (Ny et al., 1984), UK (Verde et al., 1984), and Protein C (Plutzky et al., 1986), the A and B EGF units are encoded on separate, single exons.

A particularly interesting homology is found between the low-density lipoprotein receptor (LDLR) and EGFP proteins (see Fig. 2.4) (Yamamoto et al., 1984; Russell et al., 1984; Sudhof et al., 1985). A 33 percent similarity is found over a 400aa region of the LDLR genome, which contains three repeats of the EGF B unit, corresponding exactly in location to the units d, e, and f in the EGFP gene (Yamamoto et al., 1984). In LDLR, this region is encoded by exons 7–14 of the 18 exons comprising the gene (Sudhof et al., 1985 a, b). Similarly, in EGFP, the same segment is encoded on eight exons. By comparison, of the nine pairs of introns interspersed between the exons, five pairs fall at identical positions, one pair is a few bases apart, and three pairs are unequivalent. It is therefore likely that this whole region has a common origin and

FIGURE 2.5. A diagram to show the arrangement of domains in the blood coagulation, and fibrinolytic and related proteins. Key to the diagram: Ca: Calcium binding domain; K: Kringle; EGF A: Epidermal growth factor A domain; EGF B: Epidermal growth factor B domain; FnI: Fibronectin type I domain; FnII: Fibronectin type II domain; FnIII: Fibronectin type III domain; Protease: Serine protease catalytic domain; Pro: Proline-rich region. The homologies between the domains are represented by similar shapes in the diagram. For example, the homologous Kringle and fibronectin type II domains are represented by circles. See text for details.

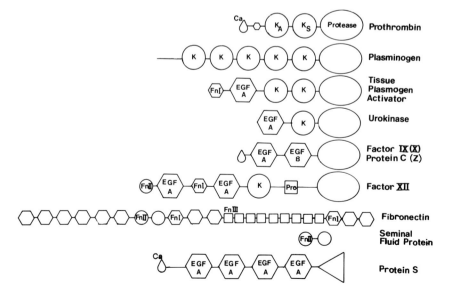

that the missing introns have been lost, or in some cases splice junction drift has occurred. Over evolutionary time it is possible that mutations in the splice region occur such that the drift of an intron boundary over some distance is possible. This will be discussed in later sections.

The binding region of the LDLR molecule is encoded by exons 2–6 and comprises seven repeats of another 40-aa sequence, of which four units occur on separate exons and three occur together on a single exon, indicating the loss of two introns in this region. This unit is repeated to create two adjacent units in the Complement 9 protein (DeScipio et al., 1984; Stanley, 1985). As the protein also contains an EGF-type unit, some distance away from the LDLR domain, it is interesting to speculate upon the gene structure, whether the individual units are separately encoded on exons, and whether the EGF unit and preceding polypeptide between the units is encoded in a similar manner to the 400-aa region separating the EGF units in the LDLR and EGF precursor molecules.

The LDLR gene thus contains 18 exons, of which 13 encode homologous segments to those found in other proteins and which also often occur on separate exons, a striking example of the correlation between protein domains and gene structure, and major evidence for exon shuffling.

The observation of a small-scale diffusion of domains, including the EGF domain, among the blood coagulation and fibrinolytic proteins, further supports the Gilbert–Blake hypothesis of exon shuffling. The cascade system of blood coagulation and fibrinolysis proceeds by conversion of a precursor serine protease to the active form by a preceding protease (Jackson and Nemerson, 1980). The catalytic domain is homologous to the pancreatic serine proteases, indicating their divergent evolution of the proteolytic domain (Young et al., 1978; Neurath, 1984). In the noncatalytic, amino-terminal segments of these proteins there are a variety of combinations of several small, disulfide-linked domains (see Fig. 2.5).

One such domain is found in Factor IX, Factor X, Protein C, Protein S, Protein Z, and prothrombin, consisting of 40 amino acids with one disulfide bridge (Hewett-Emmett et al., 1981). It also contains 10 carboxy-glutamic acid residues (Gla residues), formed in a posttranslational modification (Nelsestuen et al., 1974).

The gene structures of prothrombin, Factor IX, and Protein C indicate that the unit is encoded on an identical pair of exons (Degen et al., 1983; Macgillivray et al., 1984; Kurachi et al., 1982; Anson et al., 1984; Plutzky et al., 1986). Moreover, the eight introns of Protein C and Factor IX correspond exactly in position, indicating a recent gene duplication.

A second domain is repeated 11 times in five proteins. This domain is known as a Kringle and contains 90 amino acids linked by three disulfide bridges (Magnusson et at., 1975). The Kringle occurs five times in plasminogen, twice in prothrombin and tissue plasminogen activator, and once in urokinase and Factor XII (Magnusson et al., 1976; Claeys et al., 1976; Sottrup-Jensen et al., 1978; Malinowski et al., 1984; Pennica et al., 1983; Banyai et al., 1984; Steffens et al., 1982; Gunzler et al., 1982; McMullen and Fujikawa, 1985). In the gene structures of UK and TPA, each Kringle is encoded by a pair of exons with identical intron boundary positions (Verde et al., 1984; Ny et al., 1984). In prothrombin, a similar pair of exons encodes one Kringle, and the other is encoded on one exon (Degen et al., 1983; Macgillivray et al., 1984). The gene structures of these proteins suggest the recent evolution of the UK and TPA Kringles, which is consistent with the gene structure, whereas the comparison of the prothrombin Kringles indicates that they diverged from each other much further back in time and that an intron was either added or deleted during the subsequent evolution.

The precise definition of several exon boundaries combined with the observed autonomy of folding, structural, and functional activity of these domains strongly supports and exon-shuffling hypothesis (Trexler and Patthy, 1983; Patthy et al., 1984; Ploplis et al., 1981; van Zonneveld et al., 1986). Moreover, the argument is supported by the observation that a unit related to the Kringle is present in the multifunctional, multidomain protein, fibronectin, an extracellular matrix protein.

Fibronectin is composed of two similar chains, each containing multiple copies of three types of domain—type I, of which there are 12 copies; type II, of which there are two copies; and type III, of which there are nine copies (Petersen et al., 1983; Skorstengaard et al., 1984).

The type I domain is found in both TPA and Factor XII (Pennica et al., 1983; McMullen and Pujikawa, 1985). Only the gene structures for TPA and fibronectin are available, and these indicate that the type I unit is encoded on a single exon (Odermatt et al., 1982; Ny et al., 1984). The type II unit is found once in Factor XII and the bovine seminal fluid protein corresponds almost exactly to two repeats of the unit (Esch et al., 1983).

The type II domain has recently been shown to be related to the Kringle unit, and from comparisons of the 11 Kringle sequences with the fibronectin type II unit (Patthy et al., 1984), hydropathy studies (S. K. Holland, unpublished), and chemical structural information on the Kringle domain (Esnouf et al., 1985), it appears that the type II domain corresponds to the "core" of the Kringle unit and that the Kringle has adapted different specificities and functions in the different proteins by the addition of variable polypeptide loops on the exterior of the structure, possibly for use in anchoring the molecule at the sites of action.

Patthy has suggested that the evolutionary autonomy of the Kringle module and the relation between this and fibronectin brought about by exchange of genetic material may have faciliated further exchange and introduced the type I unit into the plasma proteins (Patthy et al., 1984). However, the relation may be complex if the observed similarity between disulfide bridges in EGF, C9, and fibronectin type I units indicates a distant homology (see Fig. 2.6) (R. F. Doolittle, 1985). This would reduce the number of domains to three—a calcium binding unit, a Kringle/FnII unit, and an EGF/FnI unit. This potential unification of domains will be discussed later in the review.

From comparisons of these protein sequences, Patthy has developed an evolutionary tree, relating each domain to the time of its insertion into the different

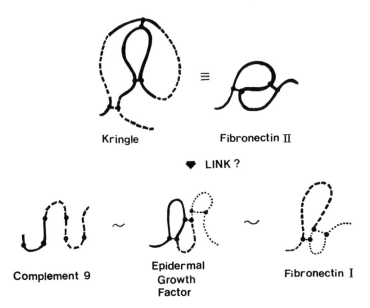

FIGURE 2.6. A diagram to show the homology between the plasma protein domains. Chain topologies are shown; disulfide bonds link the chains. Equivalent regions of the Kringle and fibronectin type II chains are shown by a solid line, ———; the regions of the Kringle that are not present in the fibronectin type II domain are shown by a dashed line, ------. The homologous regions of the Complement 9 and EGF A domains are shown by a solid line, and the homologous regions of fibronectin I and EGF A domain are shown by a dotted line, ...

proteins (see Fig 2.7) (Patthy, 1985). In their original ancestral protease, a growth factor A domain was inserted between the signal peptide and the catalytic domain. Following the divergence of the blood coagulation and fibrinolytic enzymes, a growth factor B unit was introduced into the coagulation proteases together with a calcium binding domain, through mispairing and genetic exchange. At about the same time, a Kringle domain was inserted into the fibrinolytic proteins. An ancestral, plasminogen Kringle, (Kringle S) in prothrombin was probably introduced through a double crossover, resulting in the deletion of an EGF B unit. It is notable that there is an extra loop in prothrombin (residues 48–61) containing a single disulfide bridge and corresponding to the last disulfide-linked loop of the EGF B unit. Internal multiplication of the Kringle in plasminogen occurred and the transferral of a second Kringle to prothrombin followed, the growth factor A domain being deleted from the protein by the double crossover. The growth factor A domain in plasminogen may have been lost or adapted to form the amino terminal peptide of the protein. A similar duplication of Kringles occurred in the TPA and UK ancestor, and following their divergence a fibronectin type I domain was inserted into the TPA gene. The recent determination of the Factor XII gene indicates that it is most closely related to the TPA and UK proteases and that it diverged from the

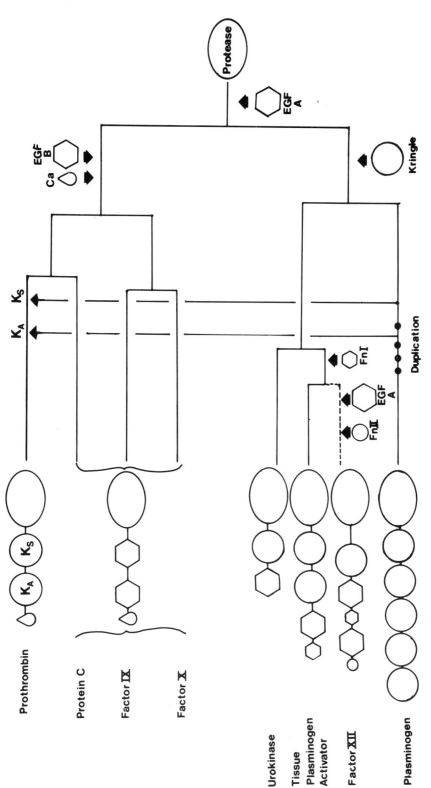

FIGURE 2.7. The proposed evolution of the blood coagulation and fibrinolytic proteins. (Adapted from Patthy, 1985.)

TPA gene after the addition of the fibronectin type I unit (McMullen and Fujikawa, 1985; Cool et al., 1985; Que et al., 1986). It is likely that a subsequent genetic exchange created a fibronectin type II unit and an EGF A domain in the Factor XII gene. The proline-rich region of this protease is simliar in some respects to the fibronectin type III unit, and the homology between the two proteins may have facilitated the insertion of a proline-rich domain into Factor XII with the loss of a Kringle. The sequence of Protein S indicates that is was evolved by gene duplication of an EGF A unit separately from both the prothrombin and Factor IX, Factor X, and protein C groups (Dahlbeck et al., 1986). In place of a protease domain it has a unique region that may be related to its involvement with the Complement system, thus linking this protein with the Complement 9 protein, which has two EGF A units (Dahlbeck and Stenflo, 1981).

The common role of these proteins is the mediation of binding of the catalyic domain to macromolecules or macroscopic structures (Jackson and Nemerson, 1980). The function of the calcium-binding domain is to bind to membranes via a calcium bridge between the gla residues and phospholipids. The pattern of residues is highly conserved in all the proteins in which this domain occurs, implying a highly defined structural involvement in binding (Young et al., 1978; Katayama et al., 1979; Fernlund and Stenflo, 1982). The other domains also contain regions of highly conserved residues in addition to the invariant cysteines. This is probably related to the structural integrity rather than to the specific function of each domain. Table 2.1 shows the range of binding functions for all the individual domains and the molecules to which they are known or proposed to bind. The specific protein–protein interactions of plasminogen may be represented by the α-antiplasmin, lysine, and arginine binding sites of the Kringles (Brockway et al., 1972; Vali and Patthy, 1982). The anchoring of the coagulation enzymes to their cofactors or membranes may be facilitated by the Kringle or EGF domains. Fibrinolytic enzymes may use fibrin as a substrate to locate their places of action. Fibronectin has many associated binding functions, using several consecutive domains to bind fibrin, actin, collagen, heparin, and DNA and possibly reflecting in this superfamily of proteins the requirement of combinations of essential elements to enable high-affinity binding to structures (Peterson et al., 1983; Skorstengaard et al., 1986). For example, the combination of an EGF and Kringle unit may be required to bind fibrin and an FnII: an EGF(A) and an FnI domain may comprise a collagen binding domain. Recent work on the TPA protein has revealed that an FnII and the second Kringle are essential for fibrin binding (van Zonneveld et al., 1986). The commonality of binding may also reflect very similar structural features in these domains and the fact that they are differentiated versions of a simple structure with subtle differences providing the specificities.

The actions of Factor IX and prothrombin demand the presence of Factor Va, which is suggested to bind to the EGF(A) unit in Factor IX and to the Kringle unit in prothrombin. This may reflect some structural equivalence in all these domains, which are all predicted to contain β-sheet, particularly about the disulfide bonds, and little α-helix. The structural integrity is clearly reliant

PROTEINS, EXONS, AND MOLECULAR EVOLUTION

Table 2.1. Protease Domain Binding Functions

Domain		Protein	Binding
Kringle		Prothrombin K(A)	Factor V?
		Prothrombin K(S)	Factor V
		Plasminogen K1	Arginine
		Plasminogen K2	Lysine
		Plasminogen K3	? Fibrin
		Plasminogen K4	Arginine
		Plasminogen K5	Lysine
		Tissue plasminogen activator (2)	Fibrin
		Urokinase (1)	?
Fibronectin II		Fibronectin (2)	Collagen
		Factor XII (1)	Collagen ?
		Seminal fluid protein (2)	?
EGF	A	Epidermal growth factor (10)	Receptor
		Tissue plasminogen activator (1)	?
		Urokinase (1)	?
		Factor XII (1)	
		Protein S(4)	
	A/B	Factor IX (2)	
		Factor X (2)	Factor V
		Protein C (2)	Factor VIII
		Protein Z (2)	
Fibronectin I		Fibronectin (5)	Fibrin
			Actin
			Heparin
		Fibronectin (3)	Fibrin
		Fibronectin (3)	Collagen
		Factor XII (1)	Fibrin ?
		Tissue plasminogen activator (1)	Fibrin

upon the use of disulfide bonds. Three-dimensional evidence for most of these proteins is unlikely for some time. However, by examining isolated domains more detail is emerging. Crystals of Fragment 1 of prothrombin, which consists of the calcium binding domain and adjacent Kringle domain, indicate that the structure comprises two distinct domains, one compact and probably the Kringle domain, containing an α-helix, and the other, long and extended, which may correspond to the calcium binding domain (Harlos, K., et al., 1987).

ANCIENT PROTEINS AND EXONS

Doolittle suggested that the split gene structure was not confined to the eukaryotes and that the genes of the earliest organisms may have contained introns (W. Doolittle, 1978). The lack of introns in prokaryotes is postulated to have arisen due to a process of removal of introns from the prokaryotes after their

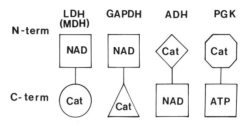

FIGURE 2.8. Diagram to show the location of the nucleotide binding domain in several proteins and its combination with catalytic domains. The abbreviations for proteins are as given in the text.

divergence from ancestral lines, in a bid to streamline the replication process. The proteins discussed earlier are relatively modern evolutionary products, evolving within the last billion years (R. Doolittle, 1985). To assess the role of the exon and the origin of the intron, we must examine the proteins of the fundamental life processes that were in existence before the separation of prokaryotes and eukaryotes.

The glycolytic enzymes have been widely studied in all cells, revealing conserved primary sequences, tertiary structures, and catalytic mechanisms. These results demonstrate their optimal evolution in the ancestor of the prokaryote and eukaryote and the lack of change in these proteins since the divergence of the cell types, about 1.5 billion years ago.

Rossmann's initial observations of a similar domain structure in the dehydrogenases led to the proposal for the divergent evolution of this domain from a common nucleotide binding ancestor. Subsequent fusion of the nucleotide binding domain with other dissimilar, catalytic domains produced the different dehydrogenases (see Fig 2.8). The domain is found in alcohol dehydrogenase (ADH), lactate dyhydrogenase (LDH), malate dehydrogenase (MDH), glyceraldehyde phosphate dehydrogenase (GAPDH), phosphoglycerate kinase (PGK), phosphorylase, and aspartate carbamoyltransferase (Eklund et al., 1976; Holbrook et al., 1975; Hill et al., 1972; Moras et al., 1975; Banks et al., 1979; Johnson et al., 1980; Monaco et al., 1978).

It was further noted that the complete nucleotide binding domain, consisting of a central, six-stranded β-sheet (βA–βF), was composed of two, similar, triple-β-stranded units about a twofold axis (Rao and Rossmann, 1973). Each unit binds either an adenine or a nicotinamide base of ATP at equivalent positions, implying the evolution of the whole domain proceeded by the combination of two "mononucleotide" binding units.

The recent determination of the gene structures of chicken GAPDH (Stone et al., 1985), human PGK (Michelson et al., 1985), and ADH (Branden et al., 1984) has enabled a comparison of the gene and protein structures. The considerable similarity in exon–intron patterns in the nucleotide binding domains of these proteins is notable, particularly in view of the correlation with elements of β-strands (see Fig 2.9).

PROTEINS, EXONS, AND MOLECULAR EVOLUTION

The nucleotide domain is encoded by five exons, the mononucleotide units separated by an intron and encoded on three and two exons, respectively. The first three exons each encode one β-strand (βA, βB, βC), whereas for the second unit, in the dehydrogenases, the first β-strand (βD) is encoded separately on one exon; the last two (βE, βF), are encoded together on one exon. In PGK the pattern is reversed. Two strands (βD, βE) are encoded on the first exon and the last one (βF) is encoded separately.

For eight of the nine introns there is direct correspondence with domain and subdomain boundaries. In ADH and GAPDH the introns occur at the boundaries between nucleotide domains and catalytic domains and between the postulated mononucleotide binding units. In PGK a similar mononucleotide interface intron occurs; it is also found at the minor crossover between nucleotide domains, although not at the major crossover.

In the enzymes GAPDH (Stone et al., 1985), pyruvate kinase (PK) (Stuart et al., 1979; Lonberg and Gilbert, 1985), and triose phosphate isomerase (TIM), (Banner et al., 1975; Straus and Gilbert, 1985), a similar correspondence between domain borders is found, with introns occurring mostly at major boundaries of domains. Also in the barrel structures of TIM and PK, the exons encode one or two α-helical or β-strand elements, indicating a distinctly nonrandom, intron distribution.

These results strongly suggest the use of several exons to encode one domain by their role in encoding folded substructures, rather than the one-exon–one-domain role, and the age of these proteins also supports the early existence of introns. In the case of the nucleotide binding domain, the evolution may have proceeded by the following route. In the ancestral gene an α–β or β–α structural element could have been encoded on one exon, (see Fig. 2.10) (Blake, 1985). Triplication of this unit would produce the mononucleotide unit. A fur-

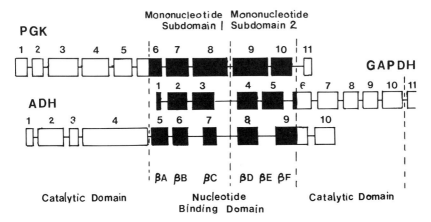

FIGURE 2.9. A diagram to show the location of introns in glyceraldehyde phosphate dehydrogenase (GAPDH); phosphoglycerate kinase (PGK), and alcohol dehydrogenase (ADH). (Adapted from Michelson et al., 1985.)

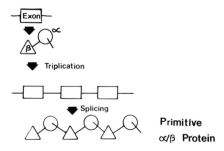

FIGURE 2.10. Proposed evolutionary pathway for the creation of the nucleotide binding unit and other primitive proteins.

ther duplication and loss of one intron would create a complete ancestral mononucleotide binding unit, encoded by five exons, providing an integral structure and useful ATP or NADH binding domain, to be utilized in different proteins by recombination with other catalytic domains (see Fig 2.8). Hence, the different classes of protein structures (Levitt and Chothia, 1976) could have been similarly created from the encodement of simple elements comprising α or β structures.

Furthermore, these proteins serve as examples of stable structures upon which different functions can be established. Structural integrity must be considered as the most important feature of a protein. Without this stability, the protein cannot reliably function either as an enzyme, involved in binding and catalysis, or as a structural molecule. Hence, in the ancient proteins, although function was a priority, the first requirements would have been to create a stable, three-dimensional structure from which functional activity could be derived and refined by point mutation in response to environmental demands. The utilization of metals in early enzymes must have been a prominent feature, whereas the inducement of allostery and structural flexibility to aid enzymic action, such as the hinge bending in hexokinase, would have developed later (Steitz et al., 1978). Similarly, the role of exons at the beginning is more likely to have been the encoding of substructures, which could then fuse together through recombination to form one domain. Go's work in correlating exons with "compact modules" of protein structure proved applicable in the prediction of intron positions in hemoglobin (Go, 1981) and has enabled similar predictions for TIM and lysozyme (Gilbert et al., 1986; Phillips et al., 1983).

Further recombination would create multidomain proteins. Subsequent point mutation at favorable functional sites is indicated by the observed separation between catalytic residues in the polypetide chain and their location on different exons. For example, on serum albumin four exons encode the functional unit of the protein (Kiousiss et al., 1981), in PGK there are five exons for the ATP binding unit (Michelson et al., 1985), and in complement Factor B, the four functional residues of the serine protease domain are carried by four separate exons (Campbell and Porter, 1983).

However, it is accepted that point mutation alone is not responsible for the observed rate of evolution on the earth. There is a large variation in protein structure, resulting in a wide range of protein function, and the observations of duplication, rearrangement, deletion, and addition of large regions of coding sequences in different proteins are simply explained in terms of an exon-shuffling mechanism to provide a rapid method of producing different proteins.

Further evidence for the use of an ancient exon-shuffling mechanism to combine different functions in one protein is found in the heme binding globins and the cytochromes. The high level of invariance of intron boundaries in species from *Xenopus* to man, particularly in myoglobin genes (separated from hemoglobin by 600–800 Myr) and the recent discovery of prokaryotic globins (Wakabayashi et al., 1986), reflects the strong requirement by the ancestral progenote for these proteins (Nishioka and Leder, 1979; Konkel et al., 1979). Each gene contains three exons that apparently code separately for heme binding ability, subunit binding, and regulation, although not for discernible structural domains (Gilbert, 1979); Blake, 1979; Eaton, 1980) (see Fig. 2.11). Re-

FIGURE 2.11. Schematic diagram of the hemoglobin subunit and its three exon products. α-Helices are labeled A–H, and the heme group is shown.

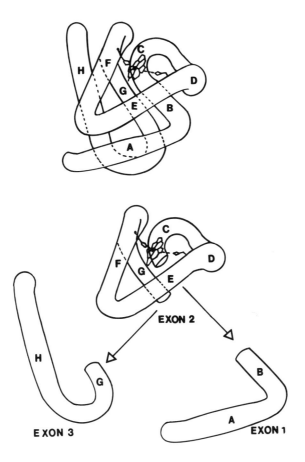

cent work has confirmed the functional independence of each of these encoded fragments and their integral role in the function of the globin (Craik et al., 1980; Craik et al., 1981). Significant correlation of chain topologies of the globins and cytochromes b_5 and c_{551} in the region of the fragment encoding heme binding with the complete polypeptides of the cytochromes indicates their divergent evolution from a common ancestral heme binding protein (Argos and Rossmann, 1979). The experimental work of Craik and deSanctis et al. (1986) verifies that the globins require the other exon-encoded fragments to enable them to bind oxygen reversibly. This suggests that following their divergence from the cytochromes, the globin genes were adapted by the addition of side exons, conferring on them reversible oxygen binding ability, and that subsequent gene duplication produced the range of globins that could then form dimers and tetramers.

The distribution and correlation of introns in the genes of large proteins containing repeated domains indicate frequently that the important mechanism of gene duplication, combined with the advantages of a split gene structure, is a common evolutionary process. Multiple exon duplication appears to be the mechanism in the evolution of the immunoglobulins, exhibiting a fivefold repeat (Sakano et al., 1979; Honjo, 1983); α-fetoprotein and its homolog, serum albumin, containing a triple repeat of four exons (Eiferman et al., 1981; Kiousiss et al., 1983; Sargent et al., 1982); ovomucoid, containing a triple repeat of two exons (Stein et al., 1980); and in particular, fibronectin (Hirano et al., 1983), thyroglobulin (Musti et al., 1986), and collagen (Wozney et al., 1981), where the number of exon repeats is as large as 40–50. The range of proteins in which this process occurs indicates that both single and multiple exons can encode structural domains in fibrous and globular proteins and that multiple exon duplication is an inherently diversifying mechanism.

Perhaps the most instructive example of multiple exon duplication is the collagen molecule, consisting of three helices, each of about 1,000 residues. The individual helices comprise a repeated structure of six adjacent tripeptides of the form Gly–X–Y (Wozney et al., 1981) (see Fig 2.12). The glycine residue at every third position is essential to the structural integrity of the collagen molecule. The tripeptide is repeated 338 times, and in the chicken α_2-procollagen gene it is encoded on 52 exons of either 54, 108, 45, or 99 base pairs in length, coding for polypeptides of $6\times$, $12\times$, $5\times$, and $11\times$ the tripeptide. By repeated duplication of the 54-base-pair exon and loss of an intron between adjacent exons, the 104-base-pair exon would be created (deCrombugghe et al., 1982). Deletion of a single tripeptide from the original 54-base-pair exon would create the forty-five-base-pair exon and, similarly, the 99-base-pair exon. Thus, the problem of the evolution of this essential structural protein is solved by a simple genetic mechanism for its construction.

THE ORIGINS OF THE MOSAIC GENE

It is important to distinguish between the role and origin of introns, noting that the gene-shuffling hypothesis relates only to possibly an incidental intron func-

PROTEINS, EXONS, AND MOLECULAR EVOLUTION

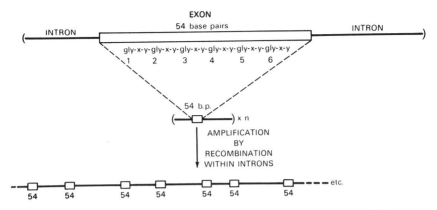

FIGURE 2.12. An evolutionary scheme for collagen. (Reproduced from de-Crombugghe and Pastan, 1982.)

tion, in response to evolutionary pressures, and not to the origin of the split gene; otherwise the evolutionary potential inherent in the theory would imply non-Darwinian, anticipatory evolution.

The substantial evidence for the ancient existence of introns, possibly even in the earliest life forms, which have subsequently been deleted in prokaryotes and partially lost in eukaryotes, appears to consolidate this view (Doolittle, 1978; Darnell, 1978; Reanney, 1979). However, there is some limited evidence to the contrary, supporting the ideas of selfish DNA and that introns were similar to transposons or mobile genetic elements, randomly inserted relatively late into the eukaryotic line by one of several methods (Cavalier-Smith, 1980; Orgel and Crick, 1980; Crick, 1979; Cavalier-Smith, 1985).

Examples of proposed intron insertion are found in the ovalbumin–α-antitrypsin (Leicht et al., 1982), actin (Blake, 1985), and serine protease genes (Rogers, 1985). In these, related genes exhibit introns that do not correspond in number of position. The opponents of intron deletion suggest that these have arisen through insertion of introns at different positions in different eukaryotic lines. This evidence appears less significant if splice junction drift by mutation is considered. The possibility of this presents a difficulty in identifying the relation between introns in different genes. Are they identical or different? Craik proposed that sliding of splice junctions may provide some evolutionary advantages, resulting in functional differences in the same protein family. As introns often map at surface loops where functions are located, mutations in the gene sequence causing insertion or deletion of amino acids may adapt the same structure for different functions (Craik et al., 1983).

Intron origin cannot be tested by nucleotide sequence analysis, as the intron sequences have been entirely randomized over the last 100 million years. The mosaic pattern is relatively stable over long periods, exemplified by the first and third introns in soybean leghemoglobin and hemoglobins that occur in identical positions in the two proteins (Jensen et al., 1981). Leghemoglobin evolved

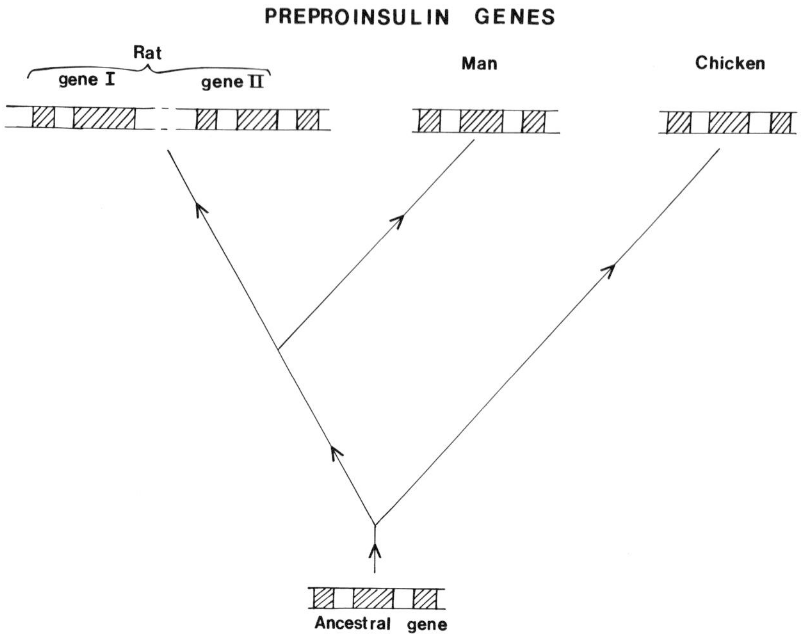

PREPROINSULIN GENES

FIGURE 2.13. The evolution of the preproinsulin gene.

prior to the β- and α-globins, an indication of the stability of the intron pattern for at least one billion years.

Evidence to support either insertion or deletion of introns would be aided by the study of the genes specifying the same protein in different species. One such example involves the genes (I and II) that encode preproinsulin in rat, as shown in Figure 2.13. Gene I has one exon and gene II has two, of which one is common to both (Lomedico et al., 1979). In gene II the intron corresponds exactly to the one in the single preproinsulin genes of man (Bell et al., 1980) and chicken (Perler et al., 1980), confirming that the ancestral gene is that containing two exons and that a subsequent gene duplication and loss of one intron in the rat genome created the structure observed today.

In the collagen gene, the deletion of an intron between two exons more simply explains the observed exon sizes than does the insertion of an intron exactly at the centers of 108-base-pair exons to produce 54-base-pair exons.

The comparison of the myosin heavy-chain (MHC) genes in rat embryonic skeletal muscle, rat cardiac muscle, and nematode muscle also suggests a differential loss of introns in different evolutionary lineages (Strehler et al., 1985). The coding sequences are strongly conserved, but the number of introns varies, the largest number occurring in the MHC embryonic gene and the least in the lower eukaryotic nematode, suggesting a highly split, common ancestral gene. The positions of several introns are correspondingly conserved, as shown in Figure 2.14. Also, the size of introns is much smaller in the nematode gene, reflecting the minimization of genome size in lower organisms. The loss of introns would appear to be nonrandom, as all the introns in the "rod" portion of the molecule have apparently been deleted, whereas those in the amino terminal region have been retained.

The gene for triose phosphate isomerase has recently been studied in chicken and maize (Straus and Gilbert, 1985; Marchionni and Gilbert, 1986). Their comparison reveals five identical intron positions, one displaced by three codons, and two extra introns in the maize gene, strongly suggesting that the ancient gene was split in the progenote before the divergence of animal and plant cells 1 billion years ago. These results strongly suggest intron loss during evolution as the exons are of regular size and one occurs in the middle of a conserved region. It is difficult to envisage intron insertion into this type of location.

In the filamentous fungi *Aspergillus nidulans* the characterized genes are both contiguous and split. In the genes for human PGK and chicken and maize TIM there are introns that correspond in position with those in the *A. nidulans* gene (Clements and Roberts, 1985; McKnight et al., 1986). The extra introns in the higher organisms reflect the likelihood of differential loss of introns in *A. nidulans*. However, in the ADH gene there is no close correlation between intron positions, although this may be indicative of intron junction drift (Branden et al., 1984; McKnight et al., 1985; Duester et al., 1986).

Despite the lack of rigorous evidence for the development of the mosaic gene, the study of the exon–intron pattern in its present form is essential to the understanding of the earliest development of life, particularly in view of the current proposals for the significant role that RNA processing played in directing precellular evolution and the development of cell lineages. If the mosaic gene existed in precellular times, then an associated splicing mechanism must also have been present (Darnell and Doolittle, 1986).

RNA processing is found in eukaryotic RNA of all types as well as in tRNA and mRNA of bacteria (Darnell, 1978). Because bacteria closely resemble the earliest organisms, this suggests the existence of a splicing mechanism in the first stages of evolution. The presence of eukaryotic RNA processing also suggests that eukaryotes are more closely related to the earliest organisms than

FIGURE 2.14. Comparison of the structures of the myosin heavy-chain genes in rat embryonic skeletal muscle (emb), rat cardiac muscle (card), and nematode unc-54 (nem). The boxes represent exons; lines represent introns. Corresponding coding positions are shown interconnected. (Adapted from Strehler et al., 1985.)

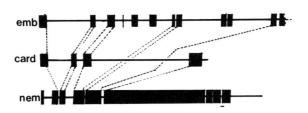

previously thought. It is more likely that both prokoryotes and eukaryotes descended from a common ancestor, the "progenote" (Woese and Fox, 1977).

Recent findings support the early theories that RNA preceded DNA, simply because of the absolute requirement by any genetic system for a translational and coding function (Crick, 1968; Orgel, 1968; Woese, 1967, 1981, 1983). This view is further strengthened by the observations of ancient introns, supporting also the necessity for RNA–RNA processing and RNA-directed development (Reanney, 1979; Reanney, 1981). The alternative theory that introns were introduced (Crick, 1979; Cavalier-Smith, 1985) appears unlikely, as a necessary prerequisite is an accurate splicing mechanism, which implies anticipatory evolution.

Evidence from bacteriophage T4 (Belfort et al., 1983; Watson et al., 1984), *Tetrahymena pyriformis* (Kruger et al., 1982; Zaug et al., 1983; Cech et al., 1983), mitochondria rRNA and mRNA from *Neurospora crassa* (van der Horst et al., 1985), and yeast mRNA (Burke et al., 1982) indicates that RNA processing, including splicing and ligation, can occur by a spontaneous, protein-free mechanism where the intron catalyzes its own excision, probably creating a three-dimensional active site from a specific base sequence at both termini. Also, the discovery of intronless pseudogenes, terminating in poly(A) tracts, implies the existence of reverse transcription from mRNA (Vanin et al., 1980). This may represent the state of affairs in precellular evolution before the divergence of cell types, as it appears unlikely that the present-day variety of sophisticated splicing mechanisms, requiring enzymes and associated factors, would be present at such an early stage of evolution. It is more likely that these represent the expansion and refinement of the initial splicing process.

RNA readily polymerizes (Van Roode et al., 1980; Inoue et al., 1983), a feature that, together with the splicing observations and other results (Stark et al., 1978; Guerrier-Takada, 1983), led RNA chemists to suggest that the primordial soup, previously defined by Haldane and Oparin (Haldane, 1929; Oparin, 1938), contained short stretches of RNA in addition to amino acids and nucleotides and that these encoded primitive but functional chains, separated by sequences now known as introns (Reanney, 1979; Doolittle, 1981; Darnell, 1983). Splicing may then not only remove the noncoding sequences, but stabilize the RNA and its transfer from the nucleus to the cytoplasm, thus enabling the first genes to be assembled and function.

Recently, the discovery of RNA enzymic activity (Cech, 1986; Westheimer, 1985) has led Gilbert to postulate that the first stage of evolution contained only RNA molecules, some of which were able to catalyze their own synthesis, possibly in a hydrothermal system (Gilbert, 1986; Nisbet, 1986). The evolution of this species of molecule proceeded by self-replication and recombination effected by the self-splicing intron. If some replicated RNA molecules spliced out their introns, they may have formed compact structures, similar to tRNA, and functioned as the first enzymes, catalyzing their own synthesis and evolving a wide range of functions and structures by recombination of exons, directed by the information storage RNA molecules that retained their introns (see Fig. 2.15).

PROTEINS, EXONS, AND MOLECULAR EVOLUTION

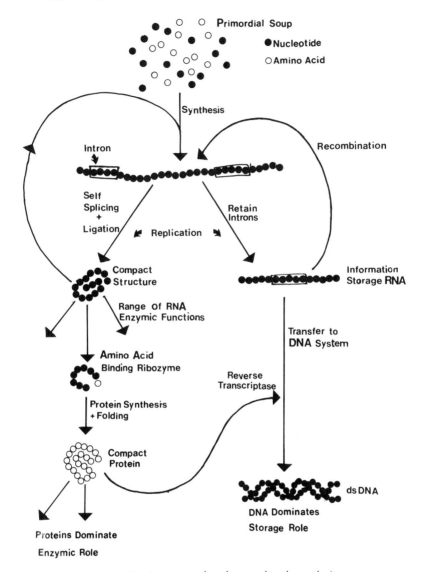

FIGURE 2.15. A proposed pathway of early evolution.

The next step in evolution would be the development of an RNA enzyme or "ribozyme" that bound amino acids and enabled them to be arranged upon an RNA template. Those combinations of amino acids that combined successfully would form the first proteins, structures that could adopt a wider variety of functions and would therefore be more effective than the ribozymes. In subsequent rounds of evolution the dominating enzyme in precellular organisms would be the protein molecule. The basis for the utilization of an exon in a ribozyme would be its ability to form a structure, and this requirement would be passed on when the first proteins were encoded. Those peptides that displayed structural ability would form the first proteins. The exons observed today reflect the selection of useful peptide structures, rather than the RNA enzymes, as these have long since disappeared and only remnants of the original functions of the RNA molecule are left.

Eventually, to provide a more stable information storage molecule, RNA would be copied into double-stranded DNA, with the aid of a reverse transcriptase protein. Hence, although the RNA molecule was the original molecule, it may have eventually delegated its two major functions to other biological molecules.

Recent work by Senapathy, when applied to RNA, comprehensively explains the origin of the segregated form of RNA into coding and noncoding regions. It also suggests why a splicing mechanism was developed at the start of primordial evolution (Senapathy, 1986). He found that the distribution of reading frame lengths in a random nucleotide sequence corresponded exactly to that for the observed distribution of eukaryotic exon sizes. These were delimited by regions containing stop signals, the messages to terminate construction of the polypeptide chain, and were thus noncoding regions or introns. The presence of a random sequence was therefore sufficient to create in the primordial ancestor the segregated form of RNA observed in the eukaryotic gene structure. Moreover, the random distribution also displays a cutoff at 600 nucleotides, which suggests that the maximum size for an early polypeptide was 200 residues, again as observed in the maximum size of the eukaryotic exon. Thus, in response to evolutionary pressures to create larger and more complex genes, the RNA fragments were joined together by a splicing mechanism that removed the introns. Hence, the early existence of both introns and RNA splicing in eukaryotes appears to be very likely from a simple statistical basis.

These results also agree with the linear relationship found between the number of exons in the gene for a particular protein and the length of the polypeptide chain (Blake, 1983). In the simplest terms, it appears that exons represented short stretches of RNA and that, under the initial selective pressures and the development of translation, those exon-encoded peptides that displayed structural ability were selected over the rest of the randomized sequences to be used as the building blocks to create larger and more useful functional proteins. The major example of this is the repeated use of the $\alpha-\beta$ unit in the ancient dehydrogenases and the barrel structures of triose phosphate isomerase and pyruvate kinase.

The actual distribution of exon size shows that most higher eukaryotic exons are of similar length of about 30–40 amino acids, with a sharp peak at about 45 residues (see Fig. 2.16). This distribution argues against a random insertion of introns (Blake, 1983). As yet there is no evidence for a mechanism of regular intron insertion, although Cavalier-Smith has proposed that the regular junctions between linkers and core particles in eukaryotic chromatin may have been hot spots for intron insertion after their divergence from the prokaryotes, which lack histones and have a more random distribution of exon sizes (Cavalier-Smith, 1985).

PROTEINS, EXONS, AND MOLECULAR EVOLUTION

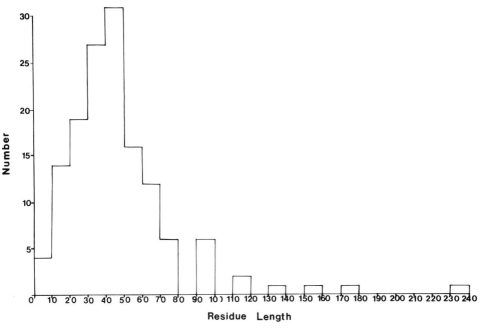

FIGURE 2.16. A histogram of exon sizes, measured by the number of encoded amino acids, in the genes of 20 different proteins.

If the legacy of intron patterns accurately reflects the ancient existence of a split gene structure in the primordial organisms, the difference in gene structures between prokaryotes and eukaryotes can be explained in terms of energy expenditure. Assuming a nonindependent evolution, following their divergence from eukaryotes, the priority for rapid growth in prokaryotes implies that the energy burden from replication, transcription, and excision of excess, unused DNA was proportionately larger than in the more slowly growing eukaryotes (W. F. Doolittle, 1978; J. E. Darnell, 1978; J. E. Darnell, Jr., 1981; W. F. Doolittle, 1981). In a bid to streamline the replication process, the prokaryotes eliminated introns. It is not inconceivable to imagine a process of intron deletion, particularly in view of the possible self-excising intron in the early stages of cellular development. It is not possible to determine whether the process of elimination is due to a deliberate deletion mechanism, recombination with DNA copies of spliced transcripts, or accidental intron loss combined with the much greater number of generations in the prokaryotic lineage. So far the preferential retention of certain introns in some eukaryotic genes and the apparently random loss of introns in others have not provided firm evidence for a specific deleting mechanism.

In eukaryotes, the development of multicellular masses in preference to faster growth appears to have been the defense against environmental challenges, a result supported by the large number of origins of replication found in eukaryotes (Newton et al., 1982; J. E. Darnell, Jr., 1981). The unused 90 percent of DNA contained in introns would consequently impose a much smaller fraction of the total energy expenditure and would allow a greater variation and expansion of the genome in developing the highly complex and differentiated organisms of the eukaryotic class. Thus, as the eukaryotes did not develop an efficient method for intron deletion, they remain as evolutionary relics.

At the next level of information, the detail encoded by the exon may reveal the fundamental aspects of structure and function that enabled certain exons to survive the first round of evolution and be utilized to create the molecules at the immediate and subsequent stages of development. As discussed previously, structural stability is vital to all proteins, and it is this feature that has been most strongly conserved thoughout evolution, unlike chemical and gene structure (Philips et al., 1983). In terms of initial functional ability, it is therefore feasible that the first, functional, exon-encoded peptides were those that could self-assemble into simple three-dimensional structures, with a significant degree of stability.

Several studies indicate that short peptides with structure and function in various environments and the existence of single, exon-encoded protein structures may have been possible at the earliest stages of life. Small polypeptides can have some defined structure in solution (e.g., ribonuclease S, which has 20 amino acids and contains 5–10 percent α-helix) (Wyckoff et al., 1970). Glucagon has 29 residues and albeit a soft protein, on binding to phospholipid it exhibits α-helical structure in a similar manner to that observed for the apolipoproteins Al and E (Sasaki et al., 1975). Recently the glucagon gene has revealed that glucagon and its homologous glucagonlike peptides are all encoded on separate exons and that they probably arose through a tandem duplication from a single glucagon coding unit (Bell et al., 1983; White et al., 1986).

Even very small peptides, such as the hormones αMSH (13 residues) and enkephalin (five residues), exhibit structure in certain environments and on interaction with receptors. It is clear that there is a variation in adopted stable conformations of short polypeptide chains. Wetlaufer showed that the smallest protein fragments with defined three-dimensional structure lie in the range of 30–50 residues, similar to the range of exon sizes in the higher eukaryotes (Wetlaufer, 1981).

These proteins appear to display a certain structural potential within an environment, and it may be this feature that is the vital link between the exon and the building of functional proteins. So far, the correlation between exons and proteins has progressed from functional domain to compact module and to substructural elements as recognized in present-day proteins. Besides the examples presented, other structural correlations with intron positions have been found in the conalbumin gene (Cochet et al., 1979), rhodopsin (Nathans et al., 1983), and β-crystallins (Blundell, 1983). However, there is no obvious universality in the approaches made so far, and despite the conservation of intron locations between species in the myosin heavy-chain genes (Karn et al., 1983) and the intermediate filaments (Marchut et al., 1984), they do not apparently delimit the observed structural domains or elements.

The selection of exon-encoded peptides with regard to their "potential" to form a structure may be the fundamental level at which the variety of exon-encoded structures are related. Depending on the degree of potential for a stable structure that each exon displayed in the surrounding environment, the exon would be utilized in different structures accordingly. For example, exons having a high structural potential and encoding very stable structures such as β-strands and α-helices would be frequently used and remain conserved throughout evolution. This appears to be validated by the observed classes of proteins containing either all α-helices, all β-sheet, alternating β and α, or mixtures of the two substructures (Levitt and Chothia, 1976). Other conformations that may be more or less favorable, depending upon the environment, would be used additionally. It is also possible that the exon-encoded structures observed today are not necessarily those present at the initial stages of evolution and that they have changed with the development of the environment. For example, in the prediction methods for secondary structures that account only for shorter-range interactions within whole proteins, it is noticeable that α and β structures are sometimes equally likely (Chou and Fasman 1974). A slight environmental change or the association with other structural units may induce the transition from one structural element to another. This also reconciles the fact that an isolated β-strand is actually a random coil, but under the correct environment, such as the vicinity of other β-stands, it has the potential and capability to form a β-strand. This may explain the observation of introns situated in the middle of α-helices or β-strands. For example, if two exons contain terminal sequences with the potential to form helices, their combination may create the correct environment for forming one helix.

Also for the domains of fibronectin and the plasma proteases, the use of similar, disulfide-linked chains may reflect some common overall potential, which upon environmental change has caused the adaptation of one structure into another, by different cysteines being linked together and subsequent deletion and insertion of other cysteines in different places. The fibronectin gene has clearly evolved by gene duplication, and a common ancestor for the FnI and FnII units appears not unlikely (see Fig 2.6). The similarity of size, predicted secondary structure content, and function of all these domains make this a feasible suggestion.

This brings us to the fundamental level of protein construction, which has so far eluded those searching for the complex code that specifies in a single polypeptide chain the information to create a three-dimensional protein. The driving force in the packing of proteins can be studied experimentally and by looking at the complete structure and assessing the variety of short- and long-range interactions that are present, but this is enormously complex. By studying the smaller exon rather than the complete protein, the fundamental interactions required for the initiation for folding may be unraveled more easily. It then may be possible to assess the way that the exon-encoded structures are packed together in the whole protein, including cooperative association between units.

CONCLUSION

From the evidence presented, it appears plausible to suppose that introns were in existence prior to the divergence of prokaryotes and eukaryotes and that they were initially present in the primodial soup, as a result of a random distribution of nucleotides creating coding regions, or exons, and noncoding regions or introns. Following the transferal of enzymic activity from RNA to protein, the exon-encoded peptides that showed structural ability were utilized as primitive enzymes and further selective pressures and recombination caused the fusion of these short polypeptides together to create larger proteins with a diverse range of functions. The first intron-splicing mechanisms were also refined, with proteins being created to carry out the process. Following the divergence of the prokaryotes and eukaryotes, in alternative survival strategies, the prokaryotes completely eliminated introns to enable faster replication at lower energy cost, and eukaryotes developed into complex, multicellular organisms taking full evolutionary advantage of the combination of structures encoded by exons, to create new proteins and diversify their lineage. So far, of the determined gene structures of present-day proteins sufficient features of the exon–intron pattern appear to have been retained to enable a provisional insight into the origins of molecular evolution. Future research into the correlation of protein structure and function with gene structure can only more closely define the nature of protein construction and the fundamental process at the center of the development of such a wide variety of molecules.

REFERENCES

Anson, D. S., Choo, K. H., Rees, D. J. G., Gianelli, P., Gould, K., Huddleston, J. A., and Brownlee, G. G. (1984). EMBO J. 3:1053–1060.

Argos, P., and Rossmann, M. G. (1979). Biochemistry 18:4951–4957.

Banks, R. D., Blake, C. C. F., Evans, P. R., Haser, R., Rice, D. W., Hardy, G. W., Merrett, M., and Philips, A. W. (1979). Nature (London) 279:773–777.

Banner, D. W., Blommer, A. C., Petsko, G. A., Philips, D. C., Pogson, C. I, and Wilson, I. A. (1975). Nature (London) 225:609–614.

Banyai, L., Varadi, A. and Patthy, L. (1983). FEBS Lett. 163:34–41.

Belfort, M., Pedersen-Lane, J., West, D., Ehrenman, K., Maley, G., Chu, F., and Maley, F. (1985). Cell 41:375–382.

Bell, G. I., Picket, R. L., Rutter, W. J., Cordel, B., Tischer, E., and Goodman, H. M. (1980). Nature (London) 284:26–32.

Belle, G. I., Sanchez-Pescador, R., Laybourn, P. J., and Najarian, R. C. (1983). Nature (London) 304:368.

Bennett, W. S., Jr., and Steitz, T. A. (1978). Proc. Natl. Acad. Sci. USA 75:4848–4852.

Blake, C. C. F. (1978). Nature (London) 273:267.

Blake, C. C. F. (1979). Nature (London) 277:598.

Blake, C. C. F. (1983). Nature (London) 306:535–537.

Blake, C. C. F. (1985). Int. Rev. Cytol. 93:149–185.
Blomquist, M. C., Hunt, L. T., and Barker, W. C. (1984). Proc. Natl. Acad. Sci. USA 81:7363–7367.
Blundell, T. (1983). Nature (London) 304:310–315.
Branden, C. I., Eklund, H., Cambillau, C., and Pryor, A. J. (1984). EMBO J. 3:1307–1310.
Breathnach, R. Mandel, J. L., and Chambon, P. (1977). Nature (London) 270:314–319.
Breathnach, R., and Chambon, P. (1981). Ann. Rev. Biochem. 50:349–383.
Brockway, W. J., and Castellino, F. J. (1972). Arch. Biochem. Biophys. 151:194–199.
Brown, J. P. Twardzik, D. R., Marquardt, M., and Todaro, G. J. (1985). Nature (London) 313:491–492.
Burke, J. M. and Rajbhandary, U. L. (1982). Cell 31:509–520.
Campbell, R. D., and Porter, R. R. (1983). Proc. Natl. Acad. Sci. USA 80:4464–4468.
Cavalier-Smith, T. (1980). Nature (London) 285:617–618.
Cavalier-Smith, T. (1985). Nature (London) 315:283–284.
Cech, T. R. (1983). Cell 34:713–716.
Cech, T. R. (1986). Cell 44:207–210.
Chous, P. Y., and Fasman, G. D. (1974). Biochemistry 13:222–245.
Claeys, H. Sottrup-Jensen, L., Zajdel, M., Petersen, T. E., and Magnusson, S. (1976). FEBS Lett. 61:20–24.
Clements, J. M., and Roberts, C. F. (1985). Curr Genet. 9:293–298.
Cochet, M., Gannor, F., Hen, R., Maroteaux, L., Perrin. F., and Chambon, P. (1979). Nature (London) 282:567–574.
Cool, D. E., Edgell, C. J. S., Louie, G. V., Zoller, M. J., Brayer, G. D., and Macgillivray, R. T. A. (1985). J. Biol. Chem. 260:13666–13676.
Craik, C. S., Buchman, S. R., and Beychok, S. (1980). Proc. Natl. Acad. Sci. USA 77:1384–1388.
Craik, C. S., Buchman, S. R., and Beychok, S. (1981). Nature (London) 291:87–89.
Craik, C. S., Rutter, W. J. and Fletterick, R. (1983). Science 220:1125–1132.
Crick, F. (1968). J. Molec. Biol. 38:367–379.
Crick, F. (1979). Science 204:264–271.
Dahlbeck, B., Lundwall, A., and Stenflo, J. (1986). Proc. Natl. Acad. Sci. USA 83:4199–4203.
Dahlbeck, B., and Stenflo, J. (1981). Proc. Natl. Acad. Sci. USA 78:2512–2516.
Darnell, J. E. (1978). Science 202:1257–1260.
Darnell, J. E. (1983). Sci. Am. 249:90–100.
Darnell, J. E., Jr. (1981). In Evolution Today: Proceedings of the 2nd International Congress of Systematic and Evolutionary Biology, Scudder, G. G. E., and Reveal, J. J., eds. Hunt Inst. for Botanical Documentation, Carnegie-Mellon University, Pittsburgh, pp. 207–213.
Darnell, J. E., and Doolittle, W. F. (1986). Proc. Natl. Acad. Sci. USA 83:1271–1275.
deCrombugghe, B., and Pastan, I. (1982). Trends Biochem. Sci. 7:11–13.
Degen, S. J. F., Macgillivray, R. T. A., and Davie, E. W. (1983). Biochemistry 22:2087–2097.
Derynck, R., Roberts, A. B., Winkler, M. E., Chen, E. Y., and Goeddel, D. V. (1984). Cell 38:287–297.

DeSanctis, G., Falcioni, G., Giandino, B., Ascoli, F., and Brunori, M. (1986). J. Molec. Biol. 188:73–76.
DeScipio, R. G., Gehring, M. R., Podack, E. R., Kan, C. C., Hugli, T. E., and Fey, G. H. (1984). Proc. Natl. Acad. Sci. USA 81:7298–7302.
Doolittle, R. F. (1984). In The Plasma Proteins, 2nd ed. Vol. 4 Putnam, S. W., ed. Academic Press, Orlando, Fla., pp.317–360.
Doolittle, R. F., Feng, D.-F. and Johnson, M. S. (1984). Nature (London) 307:558–560.
Doolittle, R. F. (1985). Sci. Am. 253:88–99.
Doolittle, R. F. (1985). Trends Biochem. Sci. 10:233–237.
Doolittle, W. F. (1978). Nature (London) 272:581–582.
Doolittle, W. F. (1981). In Evolution Today: Proceedings of the 2nd International Congress of Systematic and Evolutionary Biology. Scudder, G. G. E., and Reveal, J. J., eds. Hunt Inst. for Botanical Documentation, Carnegie-Mellon University, Pittsburgh, pp. 197–206.
Duester, G., Smith, M., Bilanchone, V. and Hatfield, G. W. (1986). J. Biol. Chem. 261:2027–2033.
Eaton, W. A. (1980) Nature (London) 284:183–185.
Eiferman, F. A., Young, P. R., Scott, R. W., and Tilghman, S. M. (1981). Nature 294:713–718.
Eklund, H. N., Nordstrom, B., Zeppezauer, E., Sonderlond, G., Ohlsson, I., Hoiwe, J., Soderburg, B. O., Tapia, O., Branden, C-I., and Akeson, A. (1976). J. Molec. Biol. 102:27–59.
Enfield, D. L., Ericsson, L. H., Fujikawa, K., Walsh, K. A., Neurath, H., and Titani, K. (1980). Biochemistry 19:659–667.
Esch, F. S., Ling, N., Bohlen, P., Ying, S. Y., and Guillernin, R. (1983). Biochem. Biophys. Res. Commun. 113:861–867.
Esnouf, P., Lawrence, M. P., Mabbutt, B. C., Patthy, L., Pluck, N. and Williams, R. J. P. (1985), Rev. Bull. Soc. Chim. Belgium 94:883–895.
Fernlund, P., and Stenflo, J. (1982). J. Biol. Chem. 257:12170–12179.
Foster, D., and Davie, E. W. (1984). Proc. Natl. Acad. Sci. USA 81:4766–4770.
Fung, M. R., Campbell, R. M., Macgillivray, T. A. (1984). Nucleic Acids Res. 12:4481–4492.
Gilbert, W. (1978). Nature (London) 271:501–502.
Gilbert, W. (1979). In ICN-UCLA Symposia on Molecular and Cellular Biology, Vol 14. Axel, R., Maniatis, T., Fox, C. F., eds. Academic Press, New York. pp.1–10.
Gilbert, W. (1986). Nature (London) 319:618.
Gilbert, W., Marchionni, M., and McKnight, G. (1986). Cell 46:151–154.
Go, K. (1981). Nature (London) 291:90–92.
Gray, A., Dull, T. J., and Ullrich, A. (1983). Nature (London) 303:722–725.
Gregory, H., and Preston, B. M. (1977). Int. J. Pept. Prot. Res 9:107–118.
Guerrier-Takada, C., Dardiner, K., Marsh, T., Pace, N., and Altman, S. (1983). Cell 35:849–857.
Gunzler, W. A., Steffens, G. J., Otting, F., Kim, S. -M. A., Frankus, E., and Flohe, L. (1982). Hoppe-Seyler's's Z. Physiol. Chem. 363:1155–1165.
Haldane, J. B. S. (1929). Rationalist Annals p. 2; reprinted in Smith, J. M., ed. (1985). On Being the Right Size and Other Essays, Oxford University Press, Oxford, pp. 101–112.

Harlos, K., Boys, C. W., Holland, S. K., Esnouf, M. P., and Blake, C. C. (1987) FEBS LEH. 224:97–103.
Hewett-Emmett, D., Czelusniak, J., and Goodman, M. (1981). Ann. NY Acad. Sci. 370:511–527.
Hill, E., Tsernoglou, D., Webb, L., and Banaszak, L. J. (1972). J. Molec. Biol. 72:577.
Holbrook, J. J., Liljas, A., Steindel, J., and Rossmann, M. G. (1975). In The Enzymes, Vol 1, 3rd ed. Boyer, P.D., ed. Academic Press, New York, pp. 191–292.
Honjo, T. (1983). Ann. Rev. Immunol. 1:499.
Hojrup, P., Jensen, M. S., and Petersen, T. E. (1985). FEBS Lett. 184:333–338.
Inoue, T., and Orgel, L. E. (1980). J. Molec. Biol. 144:579–585.
Jackson, C. M., and Nemerson, Y. (1980). Ann. Rev. Biochem. 49:727–766.
Jensen, E. O., Paludan, K., Hyldig-Nielsen, J. J., Jorgensen, P., and Marcker, K. A. (1981). Nature (London) 29:677.
Johnson, L. N., Jenkins, J. A., Wilson, K. S., Stura, E. A., and Zanotti, G. (1980). J. Molec. Biol. 140:565–580.
Karn, J., Brenner, S., and Barnett, L. (1983). Proc. Natl. Acad. Sci. USA 80:4253–4257.
Katayama, K., Ericsson, L. H., Enfield, D. L., Walsh, K. A., Neurath, H., Davie, E. W., and Titani, K. (1979). Proc. Natl. Acad. Sci. USA 76:4990–4994.
Kiousiss, D., Eiferman, F., van der Rijin, P., Gorin, M. B., Ingram, R. S., and Tilghman, S. (1981). J. Biol. Chem. 256:1960–1967.
Konkel, D. A, Maizel, J. V., and Leder, P. (1979). Cell 18:865–873.
Kruger, K., Grabowski, P. J., Zaug, A. J., Sands, J., Gottschling, D. E., and Cech, T. R. (1982). Cell 31:147–157.
Kurachi, K., and Davie, E. W. (1982). Proc. Natl. Acad. Sci. USA 79:6461–6464.
Leicht, M., Long, G. L., Chandra, T., Kurachi, K., Kidd, V. J., Mace, M., Davie, E. W., and Woo, S. L. C. (1982). Nature 297:655–659.
Levitt, M., and Chothia, C. (1976). Nature (London) 261:552–558.
Leytus, S. P., Chung, D. W., Kisiel, W., Kurachi, K. and Davie, E. W. (1984). Proc. Natl. Acad. Sci. 81:3699–3702.
Lomedico, P., Rosenthal, N., Efstratiadis, A., Gilbert, W., Kolodner, R., and Tizard, R. (1979). Cell 18:545–558.
Lonberg, N., and Gilbert, W. (1985). Cell 40:81–90.
Macgillivray, R. T. A., and Davie, E. W. (1984). Biochemistry 23:1626–1634.
Magnusson, S., Petersen, T. E., Sottrup-Jensen, L., and Claeys, H. (1975). In Proteases and Biological Control, Reich, E., Rifkin, D. B., and Shaw, E., eds. Cold Spring Harbor Laboratory, Cold Spring Harbor, N.Y. pp. 123–147.
Magnusson, S., Sottrup-Jensen, L, and Petersen, T. E. (1976). In Proteolysis and Physiological Recognition, Ribbons, D. W., and Brew, K., eds. Academic Press, New York pp. 203–238.
Malinowski, D. P., Sadler, J. E., and Davie, E. W. (1984). Biochemistry 23:4243–4250.
Marchionni, M., and Gilbert, W. (1986). Cell 46:133–141.
Marchuk, D., McCrohon, S., and Fuchs, E. (1984). Cell 39:491–498.
Marquardt, H., Hunkapiller, M. W., Hood, L. E., Twardzik, D. R., De Larco, J. E., Stephenson, J. R., and Todaro, G. J. (1983). Proc. Natl. Acad. Sci. USA 80:4684–4688.
McKnight, G. L., O'Hara, P. J., and Parker, M. L. (1986). Cell 46:143–147.

McKnight, G. L., Kato, H., Upshsall, A., Parker, M. D., Saari, G., and O'Hara, P. J. (1985). EMBO J. 4:2093–2099.
McMullen, B. A., and Fujikawa, K. (1985). J. Biol. Chem. 260:5328–5341.
Michelson, A. M., Blake, C. C. F., Evans, S. T., and Orkin, S. H. (1985). Proc. Natl. Acad. Sci. USA 82:6965–6969.
Monaco, H. L., Crawford, J. L., and Lipscomb, W. N. (1978). Proc. Natl. Acad. Sci. 75:5276–5280.
Moras, D., Olsen, K. W., Sabesan, M. N., Buehner, M., Ford, G. C., and Rossmann, M. G. (1975). J. Biol. Chem. 250:9137–9162.
Muisti, A. M., Avvedimento, E. V., Polistina, C., Ursini, V. M., Obici, S., Nitsch, L., Cocozza, S., and DiLauro, R. (1986). Proc. Natl. Acad. Sci. USA 83:323–327.
Nathans, J., and Hogness, D. S. (1983). Cell 34:807–814.
Nelsestuen, G. L., Zytovicz, T. H., and Howard, J. Biol. Chem. 249:6347–6350.
Neurath, H. (1984). Science 224:350–357.
Newton, C. S., and Burke, W. (1981). In Mechanistic Studies of DNA Replication and Genetic Recombination, ICN-UCLA Congress on Molecular and Cell Biology, Vol, 19. B. M. Alberts, ed. Academic Press, Orlando, Fla., pp.399–410.
Nisbet, E. G. (1986). Nature (London) 322:206.
Nishioka, Y., and Leder, P. (1979). Cell 18:875–897.
Ny, T., Elgh, F., and Lund, B. (1984). Proc. Natl. Acad. Sci. USA 81:5355–5359.
Odermatt, E., Tamkun, J. W., and Hynes, R. O. (1985). Proc. Natl. Acad. Sci. USA 82:6571–6575.
Oparin, A. I. (1938). The Origin of Life. Macmillan, New York.
Orgel, L., (1968). J. Molec. Biol. 38:381–393.
Orgel, L., and Crick, F. H. C. (1980). Nature (London) 284:604-607.
Patthy, L., Trexler, M., Vali, Z., Banyai, L., and Varadi, A. (1984). FEBS Lett. 171:131–136.
Patthy, L. (1985). Cell 41:657–663.
Pennica, D., Holmes, W. E., Kohr, W. J., Harkins, R. N., Vehar, G. A., Ward, C. A., Bennett, W. F., Yelverton, E., Seeburg, P. H., Heyneker, H. L., Goeddel, D. V., and Collen, D. (1983). Nature (London) 301:214–221.
Perler, F., Efstratiadis, A., Lomedico, P., Gilbert, W., Kolodner, R., and Dodgeson, J. (1980). Cell 20:555–566.
Petersen, T. E., Thogersen, H. C., Skorstengaard, K., Vibe-Pedersen, K., Sahl, P., Sottrup-Jensen, L., and Magnusson, S. (1983). Proc. Natl. Acad. Sci. USA 80:137–141.
Philips, D. C., Sternberg, M. J. E., and Sutton, B. J. (1983). In Evolution from Molecules to Men. D. S. Benndall, ed. Cambridge University Press, London and New York, pp. 145–173.
Ploplis, V. A., Strickland, D. K., Castellino, F. J. (1981) Biochemistry 20:15–21.
Plutzky, J., Hoskins, J. A., Long, G. L., and Crabtree, G. R. (1986). Proc. Natl. Acad. Sci. USA 83:546–550.
Que, B. G., and Davie, E. W. (1986). Biochemistry 25:1525–1528.
Rao, S. T., and Rossmann, M. G. (1973). J. Molec. Biol. 76:241–256.
Reanney, D. C. (1979). Nature (London) 277:598–606.
Reanney, D. C. (1981). In Evolution Today, Proceedings of the 2nd International Congress of Systematic and Evolutionary Biology. Scudder, G. G. E., and Reveal, J. J., eds. Hunt Inst. for Botanical Documentation, Carnegie-Mellon University, Pittsburgh, pp. 214–234.

Rogers, J. H. (1985). Nature (London) 315:458.
Rossmann, M. G., Moras, D., and Olsen, K. W. (1974). Nature (London) 250:194–197.
Russell, D. W., Schneider, W. J., Yamamoto,T., Luskey, K. L., Brown, M. S., and Goldstein, J. L. (1984). Cell 37:577–585.
Sakano, H., Rogers, J. H., Huppi, K., Brack, C., Traunecker, A., Maki. R., Wall, R., and Tonegawa, S. (1979). Nature (London) 277:267–623.
Sargent, T. D., Jagodzinski, L. L., Yang, M., and Bonner, J. (1981). Mol.Cell Biol. 1:871.
Sasaki, K., Dockcnills, S., Adamnak, P., Tickle, I. J., and Blundell, T. L. (1975). Nature (London) 257:751.
Scott, J., Urdea, M., Quiroga, M., Sanchez-Pescador, R., Fong, N., Selby, M., Ruther, W. J., and Bell, G. I. (1983). Science 221:236–240.
Senapathy, P. (1986) Proc. Natl. Acad. Sci. USA 83:2133–2137.
Skorstengaard, K., Thorgersen, H. C., and Petersen, T. E. (1984). Eur. J. Biol. Chem. 140:235–243.
Skorstengaard, K., Jensen, M. S., Petersen, T. E., and Magnusson, S. (1986). Eur. J. Biol. Chem. 154:15–29.
Sottrup-Jensen, Claeys, H., Zajdel, M., Petersen, T. E., and Magnusson, S. (1978). In Progress in Chemical Fibrinolysis and Thrombolysis, Vol 3. Davidson, J. F., Rowan, R. M., Samama, M. M., Desnoyers, P. C., eds. Raven Press, New York, pp. 191–209.
Stanley, K. K. (1985). EMBO. J. 4:375–385.
Stark, B. C., Kole, R., Bowman, E. T., and Altman, S. (1978). Proc. Natl. Acad. Sci. USA 75:3717–3721.
Steffens, G. J., Gunzler, W. A., Ottig, F., Frankus, E., and Flohe, L. (1982). Hoppe-Seyler's Z. Physiol. Chem. 363:1043–1058.
Stein, J. P., Caterall, J. F., Kristo, P., Means, A. R., and O'Malley, B. W. (1980). Cell 21:681–687.
Stenflo, J., and Fernlund, P. (1982). J. Biol. Chem 257:12180–12190.
Stone, E. M., Rothblum, K. N., and Schwartz, R. J. (1985). Nature 313:498–500.
Strauss, D., and Gilbert W. (1985) Molec. Cell. Biol. 5:3497–3506.
Strehler, E. E., Mahdavi, V., Periasamy, M., and Nadel-Genard, B. (1985). J. Biol. Chem. 260:468:471.
Stuart, D. I., Levine, M., Muirhead, H., and Stammers, D. K. (1979). J. Molec. Biol. 134:109–142.
Sudhof, T. C., Goldstein, J. L., Brown, M. S., and Russell, D. W. (1985a). Science 228:815–818.
Sudhof, T. C., Russell, D. W., Goldstein, J. L., Brown, M. S., Sanchez-Pescador, R., and Bell, G. I. (1985b). Science 228:893–895.
Trexler, M., and Patthy, L. (1983). Proc Natl. Acad. Sci. USA 80:2457–2461.
Vali, Z., and Patthy, L. (1982). J. Biol. Chem. 257:2104–2110.
van der Horst, G., and Tabak, H. F. (1985). Cell 40:759–766.
Vanin, E. F., Goldberg, G. I., Tucker, P. W., and Smithies, O. (1980). Nature 286:222–226.
Van Roode, J. H. G., and Orgel, L. E. (1980). J. Molec. Biol. 144:579–585.
Van Zonneveld, A. J., Veerman, H., and Pannekoek, H. (1986). Proc. Natl. Acad. Sci. USA 83:4670–4674.
Verde, P., Stopelli, M. P., Galeffi, P., Nocera, P. D., and Blasi, F. (1984). Proc. Natl. Acad. Sci. USA 81:4727–4731.

Wakabayashi, S., Matsubara, H., and Webster, D. A. (1986). Nature (London) 322:481.
Walz, D. A., Hewett Emmett, D., and Seegers, W. H. (1977). Proc. Natl. Acad. Sci. USA 74:1969–1972.
Watson, N., Gurewitz, M., Ford, J., and Spirion, D. (1984). J. Molec. Biol. 172:301–323.
Westheimer, F. H. (1986). Nature (London) 319:534.
Wetlaufer, D. B. (1981). Adv. Prot. Chem. 34:61–92.
White, J. W., and Saunders, G. F. (1986). Nucleic Acids Res. 14:4719–4730.
Woese, C. R. (1967). The Origins of the Genetic Code. Harper & Row, New York.
Woese, C. R. (1981). Sci. Am. 244:98–125.
Woese, C. R. (1983). In Evolution from Molecules to Men. Bendall, D. S., ed. Cambridge University Press, Cambridge, pp. 209–233.
Woese, C. R., and Fox, G. E. (1977). J. Molec. Evol. 9:369–371.
Wozney, J., Hanahan, D., Tate, V., Boedtker, H., and Doty, P. (1981). Nature (London) 294:129–135.
Wyckoff, H. W., Tsernoglou, D., Hanson, A. W., Knox, J. R., Lee, B., and Richards, F. M. (1970). J. Biol. Chem. 245:305–320.
Yamamoto, T., Davis, C. G., Brown, M. S., Schneider, W. J., Casey, M. L., Goldstein, J. L., and Russell, D. W. (1984). Cell 39:27–38.
Young, C. L., Barker, W. C., Tomaselli, C. M., and Dayhoff, M. O. (1978). In Atlas of Protein Sequence and Structure, Dayhoff, M. O., ed. Silver Spring, Maryland, National Biomedical Research Foundation, pp. 73–93.
Zaug, A. J. G., Grabowski, P. J., and Chech, T. R. (1983). Nature (London) 301:578–583.

3
Understanding Introns: Origins and Functions

W. FORD DOOLITTLE

Concepts of gene structure and function at the molecular level were developed in the 1960s, using *Escherichia coli* as a model, and were elaborated during the 1970s. The discovery, in 1977, that genes in complex multicellular eukaryotes are mostly made up of long stretches of noncoding DNA (introns), among which are scattered short stretches of protein-coding information (exons), challenges these concepts in several ways (Berget et al., 1977; Chow et al., 1977). This discovery requires that we explain, for instance, how it is that human genes have many introns, yeast genes have few, and *E. coli* genes have none—questions about *evolution as history*. More broadly, it asks that we redefine such notions as adaptation and function, at least as they apply to genome structure, and perhaps that we incorporate more enthusiastically into our thinking some ideas about selection, at levels other than that of individual organisms, which derive from theoretical evolutionary biology—ideas about *evolution as process*.

Gilbert's 1978 proposal concerning the "function" of intron–exon organization forms the basis for most later thinking (Gilbert, 1978, 1979). This organization speeds evolution, he argued, because information for new and potentially useful proteins, made up of already proved useful parts of old proteins, can be quickly (and reversibly) assembled through *exon shuffling*. This could occur in several ways: (1) legitimate intragenic and illegitimate intergenic recombination among exons at the DNA level, eased by the loose linkage between them; (2) inclusion at the RNA level of potentially protein-coding intronic sequences normally excluded, through mutational alteration or inherent ambiguities in splice sites within the same transcriptional unit; (3) exclusion at the RNA level of formerly exonic sequences, by the reverse of this process; and (perhaps) (4) recombination at the RNA level (through *trans*-splicing with subsequent fixation somehow into DNA) of exons originally from different transcriptional units (Solnick, 1985; Konarska et al., 1985). By these means, Gilbert reasoned: "evolution can seek new solutions without destroying the old. A classic problem is resolved: the genetic material does not have to duplicate to provide a second copy of an essential gene in order to mutate to a new function. Rather than a special duplication, the extra material is scattered in the genome, to be called into action at any time" (Gilbert, 1978).

This does not answer the question of intron origins. Implicit in Gilbert's early formulations was the assumption that introns were introduced into eukaryotic genomes at or after the time of their divergence from the genome of some ancestral prokaryote, which, like modern prokaryotes, had uninterrupted protein-coding genes (Gilbert, 1978, 1979). This way of thinking, which I will call *introns-late,* has been endorsed again recently, for instance, by Cavalier-Smith, R. F. Doolittle, and Hickey and Benkel (Cavalier-Smith, 1985; Doolittle, 1985; Hickey and Benkel, 1985, 1986). It would seem to require that introns descend from transposable elements and that selection pressures favoring their insertion be felt either at the level of the genome (in which case the elements were "selfish" [Doolittle and Sapienza, 1980; Orgel and Crick, 1980]) or at some group level above that of the individual organism, which presumably realized no immediate advantage from this invasion of the genome. The alternative, or *introns-early,* way of looking at intron origins was suggested by J. E. Darnell and by me (Darnell, 1978; Doolittle, 1978) on the basis of several considerations. The introduction of introns into already-functioning genes would, we felt, inevitably impose a serious short-term selective disadvantage on organisms, no matter what the long-term evolutionary payoff. The notion that introns must be recent because there are none in prokaryotes derives, we argued, from an incorrect view of the evolutionary relationship between prokaryotic and eukaryotic cells. Finally, we suggested that the assumption that introns never were inserted into intact, functioning genes, but were instead already present in the first cells (having been retained in eukaryotes but lost in prokaryotes) is useful in constructing scenarios for precellular genome evolution (Darnell and Doolittle, 1986).

Whatever the relative merits of the *introns-early* and *introns-late* views, it is important to realize that there are at least four good questions one can ask about intron origins and function and that separate answers to them can in fact be sought. These questions are (1) Do exons shuffle, and has exon shuffling been important in the formation of new protein-coding genes? (2) Can introns be gained or lost, and can either gain or loss alone account for their present distribution within genes and among groups of organisms? (3) What does intron distribution tell us about the evolutionary relationship between prokaryotes and eukaryotes, or conversely, what does a proper appreciation of that relationship (from other considerations) do for our understanding of introns? (4) Although it seems improper to say that introns were introduced or are maintained for the "purpose" of speeding evolution (i.e., to say that this is their *function* in any organism), is there some legitimate theoretical framework into which we can comfortably integrate the intuition that without introns we would not be here?

None of the questions has received a final answer, but we can now begin to make sensible statements about each of them.

DO EXONS SHUFFLE?

General Comments

Very shortly after Gilbert's initial publication, Blake (1978) pointed out how easily the idea of exon shuffling might be coupled with the earlier suggestions of Rossmann and others (Rossmann et al., 1974; Levine et al., 1978), that modern proteins show evidence of assembly from a limited number of globular units (structural and/or functional "domains") of ancient origin, which can be shared between proteins of different overall function. Exon shuffling would work best (or only work at all), Blake argued, if individual exons encode individual domains (or, in more recent formulations, subdomain elements of "supersecondary structure," which can be assembled in a combinatorial fashion into domains [Blake, 1983, 1985]).

It is important to understand that Gilbert's hypothesis, as extended by Blake, does not require that *all* exons encode domains. Rather, domains that are common to two proteins with otherwise independent evolutionary predigrees, and thus likely to have gotten into one or the other protein by shuffling, should be delimited by introns in both. Indeed, a kind of selection is possible here. *If* the assembly of a new protein from component parts that have already evolved toward relevant functions in other contexts is easier than the alteration of the function of existing whole proteins or the recruitment into function of random polypeptides, and *if* some such components are available because they are neatly exon encoded, then useful domain-encoding exons will be selected in the subsequent evolution of new genes. As R. F. Doolittle puts it (1985), "introns that occur between potentially useful domains will have added survival value and will predominate over those that split genes at random locations." This will be true regardless of whether introns were early or late.

Instances in which protein structure is correlated with gene structure, of which there are many (Leonard et al., 1985; Go et al., 1983; Ny et al., 1984; see following discussion) are not absolutely compelling evidence that introns were early, although they are most consistent with that view. Protein structure–gene structure correlations can still be accommodated within an introns-late interpretation that incorporates one of the following auxiliary assumptions: (1) introns were inserted randomly, but frequent shuffling and selection of the above sort have brought to the fore instances in which introns fortuitously separate domain-encoding exons, (2) introns were inserted randomly, but selection against introns that interrupt domain-encoding regions has been active at the level of RNA or protein (Craik et al., 1983; Hickey and Benkel, 1985, 1986); (3) intron insertion has been nonrandom, showing a preference, for instance, for DNA sequences likely to encode linkers between highly structured regions of proteins (D. A. Hickey, personal communication). Similarly, instances in which protein structure is *not* correlated with gene structure, of which there are also many, do not prove that introns were late. Instead, they could represent (1) the use during precellular (?) gene assembly of exons encoding unstructured polypeptides or (2) the loss of evidence for such structure through subsequent evolution (selected or neutral) of the encoded protein. A certain amount of balance in evaluating the evidence, especially in support of unitary hypotheses for intron origins, is essential.

New Genes

The human low-density lipoprotein (LDL) receptor gene provides the clearest example of exon shuffling during recent eukaryotic evolution (Sudhoff et al., 1985a,b). Most of its 18 exons define domains already identified at the protein level. The LDL-binding domain comprises seven repeats of a 40-aa sequence that is homologous to a 40-aa unit in complement factor C9, and four of these repeats are delimited by introns, at the gene level. (Two introns between repeats have apparently been lost.) Another eight exons encode a sequence homologous to part of the sequence for the precursor of epidermal growth factor (EGF), and five of the nine intron positions defining these eight exons retain introns in the EGF precursor gene. Three of the eight EGF-like exons are repeats of a sequence that is represented not only in the EGF precursor, but also (in fewer copies) in tissue plasminogen activator, urokinase, protein C, and Factors X and IX of the blood clotting system, in the gene for the last of which, at least, this repeat is also bordered by introns. Comparisons between all of these proteins, summarized recently by R. F. Doolittle (1985), reveal further sharing of still other domains, and this would appear to be but the tip of a very large iceberg that will emerge as more vertebrate protein structures and gene sequences are correlated (See Rogers [1985], for instance, for a summary of data indicating exon shuffling in the serine protease superfamily and Leonard et al. [1985] for the very recent evidence for shuffling to produce interleukin-2.)

Old Genes

Of deeper interest are older proteins, proteins that have homologs in both prokaryotes and eukaryotes and that show homologous domain structure in both. For if genes for the eukaryotic versions of these proteins have introns that separate domain-encoding exons, one must almost unavoidably conclude that these introns were instrumental in the initial assembly of the gene and that this took place before the evolutionary divergence of prokaryotes and eukaryotes, whenever in the last 4 billion years that might have occurred.

Several recent papers bear directly on this. Lonberg and Gilbert (1985) have looked at pyruvate kinase (PK). This enzyme has three well-defined domains in chicken, yeast, and bacteria. Genes for the yeast and (presumably) bacterial homologs have no introns; that for the chicken enzyme has at least 10. Nine of these interrupt the protein-coding region, dividing it into exons of roughly equal size (159 ± 47bp). This is a nonrandom distribution. Exons in this gene in general correspond to discrete secondary structural elements ("supersecondary structural units"). In the A domain, for instance, exons code for multiples of a simple supersecondary unit—an α-helix and a β-sheet. Introns do not primarily fall in regions that are variable in protein primary structure (as one might expect if their current positions reflected selection at the level of protein phenotype for minimal potential disruption [Craik et al., 1983]). Lonberg and Gilbert suggest that the gene was assembled from exons encoding short polypeptides (supersecondary structural units) and that "the non-random placement of the introns suggests that the intron-exon structure of the chicken PK is more ancient than the intronless structure of the yeast PK gene." The one domain of PK that might be shared with other enzymes, the mononucleotide binding fold region, can be aligned with its homolog in maize alcohol dehydrogenase and shares with it, at the gene level, one common intron. Glyceraldehyde phosphate dehydrogenase, the second adequately studied protein with homologs of similar structure in both prokaryotes and eukaryotes (and thus likely to predate their divergence), shows an even more striking correlation of protein structure and gene structure. Stone and co-workers (1985) have found that three of the 11 introns in the chicken gene correspond closely with the borders of the three major domains and that subdomain structure within the catalytic region is also reflected by introns in the DNA. More recently, Gilbert and others (Marchionni and Gilbert, 1986; McKnight et al., 1986) have demonstrated that triosephosphate isomerase of chicken (six introns), maize (eight introns), and *Aspergillus* (five introns) most likely evolved from a common ancestor with introns in each of the currently observed positions (plus, presumably, other positions from which introns have been lost in each of the three lineages).

These data push the origin of at least some introns back to before the divergence of prokaryotes and eukaryotes (Gilbert, 1985). If we accept Rossmann's suggestions (Rossmann et al., 1974) that the nucleotide binding domains of PK, glyceraldehyde phosphate dehydrogenase, and related enzymes arose from a common ancestor more then 3.5–4.0 billion years ago, during precellular evolution, then they push intron origins very far back indeed (Duester et al., 1986).

CAN INTRONS BE LOST OR GAINED?
The Nature of the Data

In the extreme version of the *introns-early* point of view, all introns are that old. In an extreme version of the *intron-late* view, domain-delimiting introns must be the fortuitous products of random insertions—the frequency of such fortuitous inserts perhaps being increased by the kind of selection for moveable domain-encoding exons described earlier. It is thus relevant to look within recent evolution for evidence of intron loss and intron gain.

In principle, comparison of homologous structural genes in eukaryotes of

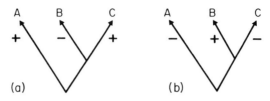

FIGURE 3.1. Phylogenetic analyses of intron loss and gain. (a) Intron present at homologous positions, in homologous genes, in species A and C, absent in B. Intron loss in lineage B is the most reasonable interpretation. (b) Intron present in B that is absent from homologous genes in A and C. Either recent acquisition in B or independent losses in A and C are reasonable interpretations.

known phylogenetic relationships can address this problem. In the example drawn in Figure 3.1a, for instance, the absence of an intron in species B that is present at homologous positions in A and C can be taken for loss in B, since its divergence from C, and presence of an intron in B that is absent from A or C can be taken for gain (Fig. 3.1b). There is a conceptual problem here, however. Independent loss of an intron from the same position in two different phylogenetic lineages (the alternative interpretation for the pattern in Fig. 3.1b) seems much more likely than independent insertion *at the same position* in two different lineages (the alternative for Fig. 3.1a). Unless this is taken into account, a likely error in such analyses is the misreading of instances of independent loss in two lineages as instances of gain in a third.

There are good examples of loss. For instance, the rat has two preproinsulin genes. One, like its homolog in chickens, has two introns; the other, only one (Perler et al., 1980). Presumably one intron was lost. In actin genes, there are at least nine positions in which introns can occur (in animals and plants; see summary in Blake et al. [1985]). No species shows introns in all nine places. There is an intron at codon 150 in chicken and rat skeletal muscle actins, human cardiac actin, and soybean actin, but not in rat β-cytoplasmic or sea urchin actins. Almost certainly there has been (independent?) loss in the last two. Similarly, an intron at codon 121 is likely to have been lost in vertebrate muscle actin.

Are there also documented instances of gain? Ueyama and co-workers find an intron between codons 84 and 85 of human smooth muscle actin, and since this is the only actin with an intron there, they ascribe it to gain (Ueyama et al., 1984). Rogers argues similarly for gain in the serine protease superfamily (Rogers, 1985). We regard this kind of evidence as equivocal (see Fig. 3.1). The discovery, for instance, of one additional phylogenetically distant actin gene with an intron between codons 84 and 85 would reverse the judgment of Ueyama et al. (1984). In the human smooth muscle actin case, the observed sporadic distribution of introns within known actins suggests that such a discovery could well be made. The analysis of mammalian α, β, and γ fibrinogen genes recently completed by Crabtree et al. (1985) seems relevant here. These authors argue that, "The more likely possibility of selective loss of introns implies that the ancestral gene, as it existed about 1 billion years ago, must have been composed of numerous small exons."

The notion that the first genes had numerous small exons, each perhaps corresponding to an element of supersecondary structure such as the $\alpha-\alpha$, $\beta-\beta$, or $\beta-\alpha$ structures of contemporary proteins, and not the larger globular domains of the initial version of the Gilbert–Blake hypothesis, is clearly gaining ground. Reworking of that hypothesis at this somewhat finer scale does no damage to the original logic (Blake, 1983, 1985). Such a notion *is* consistent with the observed sharp distribution of exon sizes around a mean length of 40–45 amino acids (Blake, 1983; 1985). The random loss of introns from a large number of ancestral sites after the divergence of two lineages could easily produce phylogenetic patterns that would mimic random insertion at a smaller number of sites, within lineages. It could also produce the poor correlation between larger-scale domain structure and intron position that has been observed for many proteins.

Selfish Introns?

The initial suggestion that introns were early rested on the intuition that the introduction of introns into once-contiguous functional genes would represent a real short-term disadvantage and no long-term advantage until many genes had been interrupted (and, furthermore, nonrandomly interrupted) so as to produce useful movable genetic modules. Even then, the advantage is *very* long term. This seems still to be true as far as selection at the level of individual organisms is concerned, but it is also now well established that self-replicating transposable elements ("selfish" DNAs) can spread throughout the genomes of a population while substantially reducing organismal fitness (up to 50 percent in the extreme case [Hickey, 1982]). Suggestions that mitochondrial introns may be transposable have thus recently been used by Cavalier-Smith to support an extreme statement of the *introns-late* position (Cavalier-Smith, 1985); also see Hickey and Benkel, 1985, 1986).

The mitochondrial data *are* tantalizing. Michel and Lang (1985) have convincingly shown that the long open reading frames of class II fungal mitochondrial introns, and of a mitochondrial plasmid, encode proteins homologous to retroviral reverse transcriptases (and their homologs in yeast and *Drosophila* transposable elements). Jacquier and Dujon (1985) and Macreadie et al. (1985) have demonstrated that the class I r1 intron in the yeast 21S rRNA gene invades uninterrupted 21S genes at high frequency, by something like gene conversion, and that this depends on the integrity of the open reading frame within the r1 element. Class I and II mitochondrial introns are, however, quite different in structure and mechanism of splicing from nuclear protein-coding gene introns, and the relevance of these observations to the more general problem is unclear. All known bona fide nuclear transposable elements generate flanking direct repeats of target site sequences upon insertion, and any element-encoded transposition enzyme that was also responsible for RNA processing (to render insertions into genes nonlethal) would need complex, multiple-sequence-recognizing abilities. An attractive transposition model based on the self-splicing rRNA of *Tetrahymena* in which circular intron RNAs insert themselves (without generating repeats) into mRNAs, which are then reverse-transcribed and reinserted (by "gene conversion"), *can* be constructed (Sharp, 1985). There is no evidence that this has ever happened, however. No general claim that introns are products of recent (since the prokaryote–eukaryote divergence) transposition can avoid having to deal with the evidence for nonrandom insertion into the DNA, at sites that are not recognizably special in the DNA, and only special in the completed protein.

INTRONS AND THE RELATIONSHIP BETWEEN PROKARYOTES AND EUKARYOTES

The *introns-late* view of intron origins emerged quite naturally from concepts of the evolutionary relationship between prokaryotes and eukaryotes that were prevalent in 1977, and it is worth recalling what those were. Stanier and van Niel had, a decade and a half before, provided a firm basis for dividing the living world into two primary kingdoms, the prokaryotes and eukaryotes, between which there was an almost unbridgeable gap—"the greatest single evolutionary discontinuity to be found in the present-day living world . . ." (Stanier and van Niel, 1962; Stanier et al., 1963). Margulis had argued, in 1967 and 1970, and others had shown by 1975, that this gap was partly bridged by plastids and mitochondria (Margulis, 1982; Gray and Doolittle, 1982). These semiautonomous organelles, whose possession was one of the fundamental features distinguishing eukaryotes from prokaryotes, were in fact prokaryotes (specifically cyanobacteria and respiring gram-negatives) that had become trapped in permanent endosymbioses within the cytoplasm of some protoeukaryotic host. The origin of that host was obscure, but it seemed only natural to think of it as descended from some third prokaryotic cell whose genome thus gave rise to modern nuclear genomes. Since large cells first appear in the fossil record some 1.4 billion years ago, this date was presumed to mark the establishment of intracellular plastids and mitochondria and thus the origin of eukaryotic cells and the nuclear genome per se. But modern eukaryotic nuclear genomes show many differences from the genomes of modern prokaryotes. These latter are of course different among themselves but still all are much more alike than any is like a eukaryotic nuclear genome. These similarities among prokaryotic genomes presumably predated the divergence of major prokaryotic lineages, between 1.5 and 2.5 billion years earlier (Knoll, 1985). Differences in genome organization and expression between modern eukaryotes and prokaryotes could thus most easily be seen as arising at or after the time of the appearance of the first eukaryotic cells. Intron insertion could logically be just one of those many differences.

The proposal that introns, although absent in prokaryotes, were nevertheless the products of precellular evolution (Darnell, 1978; Doolittle, 1978) could in this conceptual context only be taken to mean that eukaryotic nuclei had reacquired a primitive trait—a case of molecular atavism. The common sense view of genomic evolution as preceding from undefined precellular states through *E. coli*-like and yeast-like stages to modern vertebrate genomes, with increases in complexity and sophistication at every stage, was violated.

The *introns-early* view does, however, make sense in the context of an alternative view of the relationship between prokaryotes and eukaryotes that was, in 1977, just beginning to emerge from the laboratory of Carl Woese (1977). Woese and his collaborators had been using sequence information from prokaryotic 16S and eukaryotic 18S rRNAs to construct a universal phylogeny. These data clearly showed that all eukaryotes are very much more like each other than any is like any prokaryote. The reverse is also true; there is no surviving prokaryotic lineage from which the protoeukaryotic host *could* have arisen as recently as 1.5 billion years ago, even though the prokaryotic progenitors of plastids and mitochondria can clearly be identified by these methods. In fact, differences between prokaryotic 16S and eukaryotic 18S are of sufficient magnitude to suggest that, at the time of divergence of eukaryotic nuclear and prokaryotic genomes, ribosomes were still undergoing rapid evolution in response to selection for accuracy, efficiency, and speed.

A reappraisal of differences between eukaryotes and prokaryotes involving other basic components of the machineries of DNA replication, transcription, and translation supports this idea (Fox et al., 1977; Woese and Fox, 1977; Fox et al., 1980; Woese, 1982). It also supports the more general conclusion that the last common ancestor of prokaryotes and eukaryotes was a primitive cell, a cell that had a DNA genome, transcription enzymes, and ribosomes but that had as yet not solved some basic problems in genomic organization and expression. Since eukaryotes and prokaryotes descended independently from this primitive ancestor, which Woese and Fox (1977) call the *progenote,* they have solved these problems independently and differently. There is no a priori reason to regard the molecular biology of modern eukaryotes as more advanced than, or in any way derived from, the molecular biology of modern prokaryotes. Introns can in this view be understood as a primitive feature of genomes, retained in eukaryotes but lost in prokaryotes.

DO INTRONS HAVE A FUNCTION?

Evolutionary Metaphors

The questions addressed so far are historical ones: Have exons shuffled (and for how long)? Can introns be gained or lost, and if their origin predated the prokaryote–eukaryote divergence, when was that? These are profound questions, but the debate over introns also deeply involves some even more generally biological issues about the meaning of function, the nature of selection and constraints upon selection, and levels at which selection can act—questions about *evolution as a process.*

This has, in fact, been apparent throughout the debate. Gilbert's proposal was meant to address the *function* of introns ("*Why* genes in pieces?"), and he suggested that their function was to speed evolution. Darnell and I formulated the *introns-early* alternative partly because we could not see this as a function—that is, as an adaptation of organisms—produced or maintained by Darwinian natural selection simply to speed evolution. The confusion here seems to arise from the mixing of two kinds of thought. Molecular biology is firmly grounded in the neo-Darwinian paradigm, but molecular biologists often approach evolutionary questions with a kind of "commonsense" intuition that is not part of and, at least as far as language is concerned, is antithetical to that paradigm. The neo-Darwinian synthesis primarily addresses selection at the level of individual organisms (in the tradition of Mayr) or, perhaps more often, at the level of phenotype-determining genes (Mayr, 1975, 1982). Population genetics, the most highly elaborated body of evolutionary theory, deals almost exclusively with the effects on the fitness of organisms of individual genes within gene pools. Individual organisms that maintain genomic peculiarities such as introns, which in the distant future might be of use in evolving new functions, do not benefit (in terms of survival or reproduction) from doing so, and in fact will probably suffer some small selective disadvantage (Maynard-Smith, 1985). Yet, at the level of "common sense," one can construct evolutionary scenarios for introns, or for the evolution of plasticity-promoting elements in general, which have undeniable intuitive appeal. We consider three conceptually similar metaphors describing such a process.

The first was expressed best, at least for molecular biologists, by François Jacob in an article entitled "Evolution and Tinkering," published just before the discovery of introns (Jacob, 1977). Evolution does not, Jacob said, function like an engineer, constructing organisms from scratch using only the best engineering design. Instead, evolution functions like a tinkerer, using bits and pieces of whatever is at hand to construct an organism that, although not the best of all organisms, will somehow work. It is not the products of each little bit of tinkering that give us conceptual trouble—each can be seen as an expedient adaptation. It is the propensity to tinker, and the whole cumbersome genetic and developmental apparatus that makes tinkering both necessary and possible (and that may be what makes interesting things happen in evolution) that Jacob felt required explanation. The elaborate structure of theoretical population genetics is ill suited for explanations at this level.

Second, we can construct (and in fact favor) a story about the significance of introns that runs like this: Introns were present from the beginning, in the first cells, and are remnants of genome assembly in precellular evolution (Darnell, 1978; Doolittle, 1978; Darnell and Doolittle, 1985; Sharp, 1985; Cech, 1986). Introns can be lost although with difficulty, and the loss of an intron produces a small but immediate selective advantage for a single-celled organism that loses it. Other things being equal, introns will all eventually be lost, but the rate at which they will be lost has to do with the biology of the organism. In small cells that grow rapidly and expend a relatively large fraction of their total energy in making DNA and RNA (such as bacteria and yeast) intron loss represents a significant advantage. But when there is a new environment

UNDERSTANDING INTRONS: ORIGINS AND FUNCTIONS

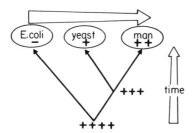

FIGURE 3.2. A scheme for intron evolution. The last common ancestor of prokaryotes and eukaryotes (the "progenote") contained many introns. Introns have been lost with time, but most rapidly in the *E. coli* lineage, less rapidly in the yeast lineage since its divergence from the human lineage, and least rapidly in the human lineage. The horizontal arrow indicates a direction of increasing morphological, developmental, and behavioral complexity.

to be exploited (and in particular to be exploited by the evolution of, for instance, a new protein that recognized in a new and complex way other cellular components and complex environmental cues), the ability to shuffle exons gives organisms a slight edge. If we look at the history of life on earth in reverse, we could see it as a process in which such complex adaptations have been increasingly at a premium. (That is, if we admit that there must have been in some sense "progress" in evolution.) So Nature has selected (at some undefinable level), at each stage, organisms with more primitive and inefficient, but for that reason more plastic, genomes, to play leading roles in the next stage in the evolution of complexity. Introns are in constant danger of being lost, but are constantly being saved by this unnamable sort of selection.

The third related metaphor can be constructed around Figure 3.2. Introns, we assume, were more prevalent at the beginning, and decrease along the vertical axis, which is time. Along the horizontal axis, in the direction of the arrow, they increase. But this horizontal axis is nothing more than a rudimentary *scala naturae,* and we of course know that no living organism is less highly evolved, or "less perfect," than any other. We have no legitimate conceptual justification for *ranking* living organisms (here in the order of both increasing complexity and increasing primitivity of genome structure—introns). Yet there remains the intuition that this way of looking at things has meaning, at some level.

As far as introns are concerned, all three metaphors convey the feeling that, although each successive new protein through shuffling is an adaptation, the process of shuffling itself is *not* an adaptation. Thus, we have difficulty even addressing this process in legitimate neo-Darwinian terms—this in face of the perception that exon shuffling may have been essential for the evolution of complex function, and in a way more important and more interesting than the gradual accumulation of adaptive mutation in structural and regulatory genes we can address with the usual neo-Darwinian language.

Adaptation, Hierarchies, and Selection

Perhaps we need new language. Gould and Vrba (1982) have recently provided what looks like the appropriate terminology. An *adaptation*, they suggest, is a character shaped by natural selection for its current use, which is its *function*. An *exaptation*, Gould and Vrba propose, is a character shaped by selection for another function (or in fact not the product of selection) that has been coopted for a new use, which is properly its *effect*. We tend to think of evolution only in terms of adaptations—traits with obvious functions in increasing the fitness of organisms that have them. The problem with introns is that they aren't like that. The use of the term *exaptation*, which recognizes that introns and many other important biological characters arose for other reasons (or for no reason) and can have effects that are not in the neo-Darwinian sense functions, gives us the freedom to talk about them in other ways.

We also may need new theory. Certainly we must recognize—as Gould, Eldredge, and others have repeatedly stressed in recent years—that there is a *hierarchy of levels* at which natural selection can act (Gould and Eldredge, 1977; Gould, 1980; 1985; Stanley, 1979; Vrba and Eldredge, 1984; Vrba, 1984). Selection, as Darwin described it, is a general principle and will operate within any *class of individuals* that are limited in time and space (that are born and die) and that show *heritable variation in fitness*—that is that (1) reproduce themselves; (2) do so in an imperfect way, with the resultant variations being themselves heritable; and (3) for which such variations affect the likelihood of survival or reproduction. Organisms are individuals of this kind. Regions of DNA whose structure or expression affect their likelihood of survival within genomes may also be individuals, and this recognition underlies the proposition that "selfish DNAs" can evolve by natural selection at a level below the usual organismal one of the neo-Darwinian synthesis (Doolittle and Sapienza, 1980; Orgel and Crick, 1980).

Ghiselin and Hull have argued that species too can be seen as individuals upon which selection can act (Ghiselin, 1975; Hull, 1976). Stanley, Vrba, Eldredge, and Gould have all discussed models for how selection at the level of species might work (Gould and Eldredge, 1977; Gould, 1980, 1985; Stanley, 1979; Vrba and Eldredge, 1984; Vrba, 1984). For "species selection" to be a distinct process, not reducible to selection at the level of organisms, it is necessary that (1) there be traits that affect rates of species reproduction (speciation) or species survival (extinction); (2) these traits not be simple extrapolations upward of traits that affect the reproduction or survival of organisms, although they can be accidental effects of such organismally adaptive traits; and (3) these traits vary heritably between lineages (clades within monophyletic groups).

It is probably best to illustrate these notions with an example (taken from Elliott Sober's book *The Nature of Selection*).

> Let's begin with two species, one winged and the other nonwinged, which initially have the same census sizes. Suppose the winged organisms survive and reproduce

UNDERSTANDING INTRONS: ORIGINS AND FUNCTIONS

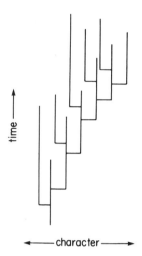

FIGURE 3.3. Schematic representation of species selection. Adapted from Sober [1984], p. 362). Character states to the right increase rates of speciation, but speciation events are random with respect to their effect on character state. Extinctions are taken to be random with time and character state.

exactly as well as their wingless counterparts. But there is a difference: small colonies of wingless individuals become separated from the main population rather frequently; winged organisms, on the other hand, in virtue of their greater mobility, very rarely form isolated subpopulations. If this system is allowed to evolve, we may later find a large number of rather small wingless species and a small number of rather large winged species. Wingless species speciate, whereas winged species grow. Notice that although wingless *species* have had more daughter species, wingless *organisms* are not more reproductively successful then winged ones.

There will thus be what is called a *trend* in the fossil record—a trend that could easily be misread as due to the cumulative effect of selection within species for organisms that are filter because they lack wings, when in fact selection has been between species, all winged and wingless organisms being as likely as individuals to survive and reproduce.

Such a trend is illustrated in another way in Figure 3.3, which is adapted from Stanley (1985) and Sober (1984). It shows a punctuated pattern of speciation, with change in some character along the *x*-axis, all changes in character state being restricted to speciation events. Speciations randomly affect this character (events are as likely to move a lineage to the left as to the right), and all species last just as long before extinction. But the rate of speciation, the frequency with which new species are born, increases with the character. There will be a trend to the right, over time. The character will have been selected for, at the level of species, because of its effect on speciation. As in the example of wingless insects, there need be no selection *within* species (i.e., between individual organisms) favoring the character. In fact, there could be selection within species against the character and yet still be an overall trend to the right, if the speciation-promoting properties of the character were sufficiently strong.

Advocates of species selection (or of Vrba's formally related "effect hypothesis") have argued repeatedly that this way of looking at evolution lies outside the modern neo-Darwinian synthesis (though not necessarily outside of Darwinism writ large). They read the modern synthesis as a story of phyletic gradualism, trends within evolution being explained by it as the continuation, across speciation events, of selection for adaptations within species (Gould and Eldredge, 1977; Gould, 1980, 1985; Stanley, 1979; Vrba and Eldredge, 1984; Vrba, 1984). Mainstream neo-Darwinists protest this as a caricature of their views (Stebbins and Ayala, 1981; Charlesworth et al., 1982). They often also disagree with species selectionists about the nature of the fossil record. For species selection to occur, species must be individuals with discrete births (speciation events) and deaths (extinctions); and for species selection to be important, most organismal adaptations must occur along with speciation—that is, evolutionary history must be primarily a history of "punctuated equilibria."

It is fortunately not necessary to take sides on all these issues to agree with Sober (1984), who writes that

> . . . species selection is a very different sort of evolutionary mechanism from individual selection. This is the important sense in which the ideas collected under the banner of punctuated equilibrium effectively "decouple" macro- and micro-evolution. This point goes beyond the fairly obvious idea that macrophenomena may be studied "at their own level," rather than from a micro point of view. That is a relatively uncontroversial epistemological or methodological thesis. Rather, we have here an ontological claim to the effect that an item at the macro-level is not identical with an item at the micro-level.

Selection Against Introns

Although some introns in some eukaryotic protein-coding genes now serve regulatory roles, so that tissue-specific differential splicing may sometimes be important in development (Breitbart et al., 1985; Nawa et al., 1984), the bulk of the data suggest that most introns do not contribute to the fitness of organisms that bear them, in terms of survival or reproduction. Within species (during phyletic evolution), introns should be selected against to the extent that their replication, transcription, and processing are burdensome. Intron elimination may, of course, be painfully slow, since (1) there seems to be no mechanism for the concerted elimination of numerous introns from numerous genes simultaneously; (2) in genomes with many introns, the loss of only one or a few may represent a miniscule selective advantage; and (3) elimination requires precise deletion, perhaps most often by a pathway resembling processed pseudogene formation. Nevertheless, we argue, if selection against introns within species is sufficiently strong (as in bacteria and simple eukaryotes specializing in small size and rapid growth and development), then any selection for introns between species can be overcome.

Overall DNA content (C-value) generally correlates well with organismal or cellular size and generation time or developmental rate, and some correlation between C-value and intron content might be expected in any case (Cavalier-Smith, 1985). We can make a more specific prediction, however. If the absence of introns in modern prokaryotes is to be explained by loss due to selection for efficiency in genome organization and expression ("genomic streamlining"), and this explanation is generalizable, then there should also be within eukaryotes a correlation between C-value and *number* of introns per unit length of DNA. A relationship between C-value and the relative *amount* of DNA in introns is also expected, but would not by itself provide support for any specific hypothesis concerning rates of loss of introns in prokaryotes and eukaryotes.

Introns and Evolutionary Constraints

Progressive intron loss means progressive loss in the ability to evolve new function by exon shuffling. To the extent that exon shuffling can be important in organismal evolution, then, the evolutionary potential of a group of organisms will be *constrained* (both limited and in a sense directed) by the complement of protein-domain-defining introns its genomes fortuitously retain. Evolutionary biologists interested primarily in development have recently begun to look again seriously at the notion that constraints—biases "on the production of variant phenotypes or limitation(s) on phenotype variability caused by the structure, character, composition or dynamics of the developmental system" (Maynard Smith et al., 1985)—may be as important as adaptation in determining the course of organismal evolution. They recognize that constraints may also exist at the level of the genotype. Maynard Smith et al. (1985), for instance, note that

> Evolutionary transitions from one (adult) phenotype to another are mediated by genetic change. Suppose that a utopian analysis of the selective regime faced by a given lineage in a particular environment were to reveal that a certain evolutionary transition would be likely if a particular sequence of phenotypes, each selectively preferable to its predecessor, were available. Even in this ideal case, it would not follow that the lineage would make the transition in question—for the requisite variation to produce the desired sequence of phenotypes (and to do so in the proper order) might not be available. There *must* be a genetic pathway, i.e., a series of mutations, accessible to the lineage, for it to produce the phenotypes of the sequence in the proper order. But the accessibility of such a pathway is not guaranteed.

In this conceptual framework we might suggest, for instance, that had one known introns were distributed in the genomes of vertebrates before the origin of the more recent components of the blood–protein system, one might have predicted (with some probability of success) the evolutionary development of that system and, given comparable knowledge, perhaps predicted a different course for the evolution of invertebrate blood systems. On a larger scale, one could, for example, begin to address in a more precise fashion the suggestion that prokaryotes and eukaryotes evolve in qualitatively different ways (Gilbert,

1978; Knoll, 1985; Carlile, 1980). Just how useful or important such retrospective analysis could be remains uncertain, but the relevant data base is emerging rapidly.

Selection for Introns

Evolutionary lineages that retain introns, then, have greater evolutionary potential. Introns, like transposable elements, mutation, sex, and recombination, can "speed evolution." But as Maynard Smith et al. (1985) recently noted, in discussion how the latter might arise or be maintained by group or species selection,

> A major dificulty stands in the way of this approach to the plasticity and directive power of the genome. It stems from the fact that selection between populations or species is likely to be a weak force compared to selection between individuals or genes. If genetic structures are to evolve because they promote evolution, this "weakness" of selection between populations and species must be compensated by the cumulative advantageous effect of the operation of those structures over long spans of evolutionary time. . . .

I would like to suggest that, for introns, the "weakness" of selection above the level of individual organisms has been indeed at least partially compensated by the cumulative advantageous effect over long spans of evolutionary time, these effects having been realized roughly in proportion to the degree to which introns have been retained. That is, I suggest that the present distribution of introns, metaphorically represented in Figure 3.2, can be understood in terms of a species-selection-type model that would incorporate the following considerations: (1) Useful intron-facilitated evolutionary changes—exon shuffles—are likely to occur less frequently than populations within species, or even species themselves, arise and go extinct. (2) Thus, any accounting for intron maintenance invoking selection must address selection at the level of species, at least. (3) Major taxa that retain relatively many introns will, other things being equal, show higher rates of speciation, or lower rates of extinction than taxa that retain relatively few introns. (4) Higher rates of speciation could reflect nothing more than more frequent fixation, within peripheral isolates, of useful new genes arising through exon shuffling, and subsequent divergence (through drift or selection) of those isolated populations from the parental populations. No novel genetic mechanism for speciation itself need be invoked, and allopatry might still be the major route to species formation (Mayr, 1975, 1982; Wright, 1982; Mary, 1963; Lewontin, 1974). (5) This notwithstanding, novel adaptations acquired by shuffling might in general more often lead to "quantum" differences in phenotype. One would expect that in general new enzymatic or other physiological activities acquired by point mutational change will differ only quantitatively from preexisting activities, whereas new, complex, coordinated activities assembled in a combinatorial fashion from preexisting separate component activities might present "qualitative" differences. (6) The relative retention of introns in complex, multicellular eukaryotes (as compared with simpler eukaryotes, or prokaryotes) can thus be understood as due to a species-level *function* of introns that arises from an organism-level *effect*. That is, introns are adaptations for species because they promote more frequent successful adaptations of peripheral isolates of organisms to novel niches and thus increase rates of speciation in groups that have retained them—producing a trend toward the (relative) retention of introns (see Figs. 3.2 and 3.3). (7) If, furthermore, most changes in organismal phenotype occur at times of speciation, as advocates of "punctuated equilibria" claim, then it is also *indirectly* the case that introns are retained because they facilitate organismal adaptation (although not generally within species during "phyletic evolution"). (8) It is of course not the case that individual introns, as independent genetic entities, are separately maintained by selection by this level. One could not argue, for instance, that selection had specifically "saved" an intron separating domain-encoding exons that have never been shuffled since the origin of cells. Rather, the propensity to retain introns must be seen as lineage-specific character, related to some general way to those molecular, physiological, developmental, and ecological factors that promote or permit high C-values and the retention of other forms of "selfish" or "junk" DNA, all of which may contribute to the generation of evolutionary novelty. (9) Selection at the level of species can at least partially account for the relative retention (the *maintenance*) of introns in certain groups. We doubt that such selection can account for intron *origins*. A similar uncoupling of selective pressures and mechanisms involved in origins and maintenance has been invoked in discussions of the evolution of transposable elements, mutation rates, sexual reproduction, and genetic recombination. Vrba, and Vrba and Eldredge have quite reasonably criticized earlier formulations of species-selection models for failing to distinguish speciation-promoting characters that are emergent at the level of species from those characters that are "simple sums of organismal adaptations" (Vrba, 1980; Vrba and Eldredge, 1984; Vrba, 1984). Whether the model for understanding introns I propose here is similarly subject to that criticism depends upon (1) whether speciation (or extinction) is something that species, rather than organisms, do; (2) how tightly coupled important adaptive shuffling and speciation events are in time or mechanism and more generally, (3) how we are to understand cause-and-effect relationships between adaptation and reproductive isolation during speciation (Wright, 1982; Mayr, 1963; Lewontin, 1974).

The Meaning of Function

In a certain sense, then, I agree with Gilbert (1979) that introns "speed evolution" and with R. F. Doolittle that shuffling "must have selective value in exploiting new situations." And in that sense, what I have tried to point out in this essay is that revolutionary arguments for the utility of introns made by molecular biologists, on the basis of the kind of intuitive common sense described earlier, may be correct, *but* they are not straightforward extrapolations upward of the kind of mainstream neo-Darwinism in which theoretical molecular biology has hitherto been embedded, and they are not couched in language that adequately expresses that disconnection. Such arguments, to be experimentally useful and conceptually manipulatable, must explicitly recognize that life and evolution are heirarchical, that there are many levels at which selection can act, and that it often does so in opposite directions (in terms of some particular trait) at different levels. Organismal evolution is the complex result of these multilevel interactions.

REFERENCES

Berget, S. M., Moore, C., and Sharp, P. A. (1977). Proc. Natl. Acad. Sci. USA 74:3171.
Blake, C. C. F., (1978). Nature 273:267.
Blake, C. C. F. (1983). Nature 306:535.
Blake, C. C. F. (1985). Int. Rev. Cytol. 93:149.
Breitbart, R. E., Nguyen, H. T., Medford, R. M., Destree, A. T., Mahdavi, V. and Nadal-Ginard, B. (1985). Cell 541:67.
Carlile, M. J. (1980). Symp. Soc. Gen. Microbiol. 30:1.
Cavalier-Smith, T. (1985). Nature 315:283.
Cavalier-Smith, T. (1985). The Evolution of Genome Size. Wiley, New York.
Cech, T. R. (1986) Proc. Natl. Acad. Sci. USA 83:4360–4364.
Charlesworth, B., Lande, R., and Slatkin, M. (1982). Evolution 36:474.
Chow, L. T., Gelinas, R. E., Broker, T. R., and Roberts R. T. (1977). Cell 12:1.
Crabtree, G. R., Comeau, C. M., Fowlkes, D. M., Fornace, A. J., Jr. Malley, J. D., and Kant, J. A. (1985). J. Molec. Biol. 185:1.
Craik, C. S. Rutter, W. J., and Fletterick, R. (1983). Science 220:1125.
Darnell, J. E., Jr. (1978). Science 202:1257.
Darnell, J. E., Jr., and Doolittle, W. F. (1986). Proc. Natl. Acad. Sci. USA 83:1271–1275.
Doolittle, R. F. (1985). Trends Biochem. Sci. 10:233.
Doolittle. W. F. (1978). Nature 272:581.
Doolittle, W. F., and Sapienza, C. (1980). Nature 284:601.
Duester, G. Jornvall, Y., and Hatfield, G. W. (1986). Nucleic Acids Res. 14:1931–1941.
Fox, G. E., Magrum, L. J., Balch, W. E., Wolfe, R. S. and Woese, C. R. (1977). Proc. Natl. Acad. Sci. USA 74:4537.
Fox, G. E., Stackebrandt, E., Hespell, R. B., Gibson, J., Maniloff, J., Dyer, T. A., Wolfe, R. S., Balch, W. E., Tanner, R. S., Magrum, L. J., Zablen, L. B., Blakemore, R., Gupta, R. Bonen, L., Lewis, B. J., Stahl, D. A., Luehrsen, K. R., Chen, K. N., and Woese, C. R. (1980). Science 209:457.
Ghiselin, M. T. (1975). Syst. Zool. 23:536.
Gilbert, W. (1978). Nature 271:501.
Gilbert, W. (1979). In Eukaryotic Gene Regulation: ICN-UCLA Symposia on Molecular and Cellular Biology Vol. 14. Axel, R., Maniatis, T., Fox, C. F., eds. Academic Press, New York, p. 1.
Gilbert, W. (1985). Science 228:823.

Go, M. (1983). Proc. Natl. Acad. Sci. USA 80:1964.
Gould, S. J. (1980). Paleobiology. 6:119.
Gould. S. J. (1985). Paleobiology 11:2.
Gould, S. J., and Eldredge, N. (1977). Paleobiology 3:115.
Gould, S. J., and Vrba, E. (1982). Paleobiology 8:4.
Gray, M. W., and Doolittle, W. F. (1982). Microbiol. Rev. 46:1.
Hickey, D. A. (1982). Genetics 101:519.
Hickey, D. A., and Benkel, B. F. (1986). J. Theor. Biol. 121:283–2911.
Hull, D. L. (1976). Syst. Zool. 25:174.
Jacob, F. (1977). Science 196:1161.
Jacquier, A., and Dujon, B. (1985). Cell 41:383.
Knoll, A. H. (1985). Paleobiology 11:53.
Konarska, M. M., Padgett, R. A., and Sharp, P. A. (1985). Cell 42:165.
Leonard, W. J., Depper, J. M., Kanelisa, M., Kronke, M., Peffer, N. J., Svetlik, P. B., Sullivan, M., and Greene, W. C. (1985) Science 230:633.
Levine, M., Muirhead, H., Stammers, D. K., and Stuart, D. I. (1978). Nature 271:626.
Lewontin, R. C. (1974). The Genetic Basis of Evolutionary Change. Columbia University Press, New York.
Lonberg, N., and Gilbert, W. (1985). Cell 40:81.
Macreadie, I. G., Scott, R. M., Zimm, A. R., and Butow, R. A. (1985). Cell 41:395.
Marchionni, M., and Gilbert, W. (1986). Cell 46:133–141.
Margulis, L. (1982). Origin of Eukaryotic Cells. Yale University Press, New Haven.
Maynard Smith, J. (1978). The Evolution of Sex. Cambridge University Press, Cambridge.
Maynard Smith, J., Burian, R., Kauffman, S., Alberch, P., Campbell, J., Goodwin, B., Lande, R., Raup, D., and Wolpert, L. (1985). Quart. Rev. Biol. 60:265.
Mayr, E. (1963). Animal Species and Evolution. Harvard University Press, Cambridge, Mass.
Mayr, E. (1975). Biol. Zentralbl. 94:377.
Mayr, E. (1982). The Growth of Biological Thought. Harvard University Press, Cambridge, Mass.
McKnight, G. L., O'Hara, P. J., and Parker, M. L. (1986). Cell 46:143–147.
Michel, F., and Lang, R. F. (1985). Nature 316:641.
Nawa, H., Kotani, H., and Nakanishi, S. (1984). Nature 312:729.
Ny, T., Elgh, F., and Lund, B. (1984). Proc. Natl. Acad. Sci. USA 481:5355.
Orgel, L. E., and Crick, F. H. C. (1980) Nature 284:604.
Perler, F., Efstratiadis, A. A., Lomedico, P., Gilbert, W., Kolodner, R., and Dodgson, J. (1980). Cell 20:555.
Rogers, J. (1985). Nature 315:458.
Rossmann, M. G., Moras, D., and Olsen, K. W. (1974). Nature 250:194.
Sharp, P. A. (1985). Cell 42:397.
Sober, E. (1984). The Nature of Selection: Evolutionary Theory in Philosophical Focus. MIT Press, Cambridge, Mass.
Solnick, D. (1985). Cell 42:157.
Stanier, R., Adelberg, E. A., and Douderoff, M. (1963). The Microbial World. Prentice-Hall, Englewood Cliffs, N.J.
Stanier, R.Y., and van Niel, C. B. (1962) Arch. Mikrobiol. 42:17.
Stanley, S. Macroevolution: Pattern and Process. W. H. Freeman, San Francisco, 1979.
Stanley, S. (1985). Paleobiology 11:13.
Stebbins, G. L., and Ayala F. J. (1981) . Science 213:967.

Stone, E. M., Rothblum, K. N., and Schwartz, R. J. (1985). Nature 313:498.
Sudhoff, T. C., Goldstein, J. L., Brown, M. S., and Russell, D. W. (1985a). Science 228:815.
Sudhoff, T. C., Russell, D. W., Goldstein, J. L., Brown, M.S., Sanchez-Pescador, R., and Bell, G. I. (1985b). Science 228:893.
Ueyama, H., Hamada, H., Battula, N., and Kakunaga, T. (1984). Molec. Cell. Biol. 4:1073.
Vrba, E. (1984). Syst. Zool. 33:318.
Vrba, E., and Eldredge, N. (1984). Paleobiology 10:146.
Woese, C. R. (1982). Zbl. Bakt. Hyg. I. Abt. Orig. C3:1.
Woese, C. R., and Fox, G. E. (1977). Proc. Natl. Acad. Sci. USA 74:5088.
Wright, S. (1982). Evolution 36:427.

4
Intron-Dependent Evolution of Progenotic Enzymes

EDWIN M. STONE AND ROBERT J. SCHWARTZ

The purpose of this volume is to discuss the biology of introns—interruptions in the orderly storage of genetic information that were unknown before 1978, that were considered eukaryotic curiosities for several years thereafter, and that are now slowly being recognized as important participants in the evolution and regulation of the genes they interrupt. In this chapter we summarize evidence that supports the idea that introns and RNA splicing facilitated the evolution of a series of related enzymes. We also show that the same evidence that supports an evolutionary role for introns of genes shared by prokaryotes and eukaryotes (termed *progenotic genes* in this chapter) also supports the notion of Doolittle (1978) that introns and RNA splicing are evolutionarily ancient, present before the construction of some of the oldest known genes. We suggest that several different types of evolutionary events probably gave rise to the numerous introns found in modern vertebrate genes and that no single explanation need be put forward to account for all of them. Lastly, we propose a mechanism, dependent upon the existence of RNA splicing, whereby a functioning enzyme could evolve *before* its gene.

In the mid-1970s, before the discovery of intervening sequences, the prevailing concept of the ontogeny of eukaryotes was that they had evolved from a symbiotic fusion of simple, prokaryote-like ancestral organisms. The vast differences between eukaryotes and prokaryotes in DNA content, chromatin structure, and gene regulation were thought to result from stepwise, evolutionarily advantageous increases in complexity, from that of one prokaryote embedded in the cytoplasm of another (the first nucleus) to that of a modern eukaryotic nucleus (Margulis, 1970). Thus, when intervening sequences were discovered to interrupt the coding sequence of eukaryotic genes, the simplest explanation was that this represented yet another *addition* to the structure and function of the eukaryotic nucleus in its evolution from a prokaryotic ancestor. In this context, several investigators pointed out potential evolutionary advantages for organisms bearing introns and RNA splicing machinery and suggested that these advantages might have been of sufficient magnitude to warrant the evolutionary development of such a complicated gene organization. Chief among these were Walter Gilbert, Ford Doolittle, and Colin Blake.

Gilbert (1978) suggested that the presence of introns would offer several evolutionary advantages to a eukaryotic organism possessing them.

1. Point mutations (occurring at splice junctions) could effect large insertions or deletions in a gene's protein product, causing much greater change than the mutation of a single amino acid that would occur in a nonmosaic gene.
2. Such a splice junction mutation might be incompletely expressed, so that the original gene product *and* the mutant product could be expressed simultaneously at significant rates, allowing an organism to experiment with gene variations without first duplicating the parent gene (whose mutation might otherwise be lethal).
3. If more than one useful product could be made from a single DNA sequence (by differential splicing), the splicing pattern could be modulated during cell differentiation by another molecule (e.g., a different splicing enzyme).
4. If an exon, or group of exons, encoded a peptide with a given function, unequal recombination within the bordering introns could translocate that function as a unit.

Doolittle (1978) expanded these ideas in an important way. He pointed out that it was difficult to envision an evolutionary advantage in dividing previously intact genes. He suggested that intervening sequences and RNA splicing were not recent evolutionary developments restricted to modern eukaryotes, but that they were vestiges of the gene organization of the oldest organisms. He argued that the compact gene organization of modern prokaryotes reflects an evolutionarily advantageous *loss* of introns and RNA splicing. This view did not invalidate the arguments of Gilbert in any way; it simply extended them into the realm of the primordial organisms and the earliest genes.

Just as important as the early theories about the origin and function of intervening sequences was the thinking about ways in which these new ideas could be supported experimentally. Blake (1978) outlined a mechanism for demonstrating intron-dependent evolution of genes when he pointed out that most larger proteins are composed of independently folded domains that could easily be the result of the association (at the DNA level) of previously separate gene segments (exons). That is, if certain exons encoded peptides that were capable of a useful function in more than one parent protein (see Gilbert's idea 4 preceding), then the structure of the peptide should be discernibly independent from the structure of the remainder of the parent proteins. Thus, gene structure might be expected to be mirrored in protein structure, and any demonstration of a protein domain with a known biological function, encoded by a gene segment delimited by introns, would be suggestive evidence that shuffling exons from gene to gene could result in shuffling useful functional domains from protein to protein.

As discussed extensively by Holland and Blake in chapter 2 of this volume, the evidence for intron-mediated evolution of *eukaryotic* genes is now overwhelming. The domains of several proteins have been shown to be encoded by exons or groups of exons (Stein et al., 1980; Barker et al., 1980; Tilghman, 1982). The most striking example is probably that of the LDL receptor gene (Sudhof et al., 1985), in which the functional domains have been shown to be recruited from other known genes (EGF, clotting factor IX, and complement factor C9). However, these results do not speak to the question of RNA splicing in primordial organisms. That is, the demonstration of intron-mediated evolution of genes found only in eukaryotes is completely consistent with the initial view that introns arose *after* the divergence of prokaryotes and eukaryotes. To examine Doolittle's idea of ancient RNA splicing, one must look at genes shared by prokaryotes and eukaryotes. Demonstration of intron-mediated evolution of such a shared gene would strongly support Doolittle's idea.

Figure 4.1 may clarify the mechanism of such an argument. The first assumption is that there was, in fact, a common ancestor of modern prokaryotes and eukaryotes. Inasmuch as these lines share a common genetic code, it is reasonable to assume that protein translation from nucleic acid genes existed in a primordial organism that later gave rise to both the prokaryotes and the eukaryotes. Woese and Fox (1977) call such an organism the *progenote*. The next element in the argument is the identification of some protein or proteins (e.g., glycolytic enzymes) that were present in the progenote and that have been conserved in modern prokaryotes and eukaryotes. The latter condition is necessary if we are to be able to make inferences about the structure of the protein in the progenote based on the structure of the modern versions. In addition, conservation of a protein across the evolutionary distance separating prokaryotes and eukaryotes establishes, in and of itself, the presence of the protein in the progenote. This is shown by comparing the two possibilities in Figure 4.1. In Figure 4.1A the ancestor did not possess glycolysis, and prokaryotes and eukaryotes developed this pathway independently. In Figure 4.1B the ancestor possessed glycolytic enzyme genes that were conserved during the divergence of the two daughter lines.

As already mentioned, evidence to support the latter consists of comparisons of the amino acid sequence of glycolytic enzymes from modern prokaryotes and eukaryotes. Figure 4.2 demonstrates the extreme conservation of GAPDH in eight species: two bacteria, yeast, an invertebrate, and three vertebrates (Stone et al., 1985a). With 20^{332} sequence possibilities for this 36-kDa enzyme it seems extraordinarily unlikely that this amino acid homology reflects an *apparent* conservation that in reality arose from convergent evolution. It is much more likely that modern prokaryotes and eukaryotes evolved from an ancestral organism that was equipped with several glycolytic enzymes that were essentially identical in structure and function to those present in both prokaryotes and eukaryotes today. To put it another way, the amino acid homology between

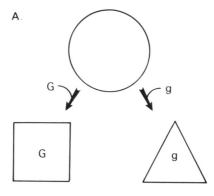

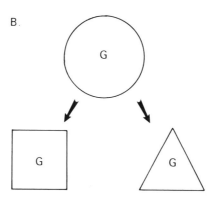

FIGURE 4.1. Two possibilities for the evolutionary origin of the glycolytic enzymes of prokaryotes and eukaryotes. In A the progenote (circle) lacks glycolytic enzymes (G,g) that evolve independently during the divergence of prokaryotes (square) and eukaryotes (triangle). In B the progenote contains glycolytic enzymes (G) that are inherited by prokaryotes and eukaryotes.

glycolytic enzymes of prokaryotes and eukaryotes suggests that the evolutionary assembly and refinement of several major glycolytic enzymes was essentially complete in the progenote.

The last step of the argument concerns the presence of RNA splicing in the progenote. If RNA splicing facilitated the evolution of the glycolytic enzyme genes present in the progenote, RNA splicing had to be present in the progenote. Thus, when the same type of demonstrations used to support intron-mediated evolution in eukaryotic genes are applied to glycolytic enzyme genes (and other related progenotic genes), the result is the support of Doolittle's contention that RNA splicing was present in the progenote.

To restate, then, the requirements for a gene–protein pair to be useful for

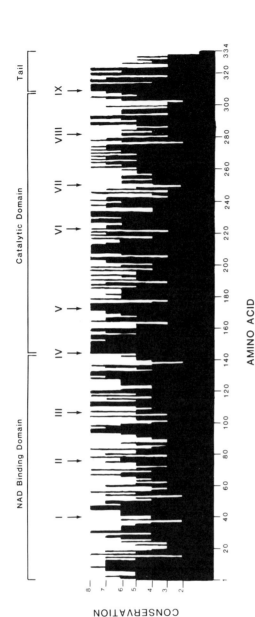

FIGURE 4.2. Conservation of GAPDH amino acid sequences among seven species (eight genes). Amino acid sequences from two bacteria, yeast (two genes), lobster, pig, chicken, and human were aligned and compared for sequence homology. Occasionally, insertions or deletions of an amino acid were assumed to maintain the sequences in register, but in no case was the number of assumptions per sequence greater than three. The amino acid residue numbers of the composite GAPDH protein are given on the abcscissa and the number of genes that agree at each position is plotted on the ordinate. Roman numerals indicate the positions of intervening sequences in the chicken GAPDH gene. (From Stone et al., 1985a.)

studying intron-dependent evolution in the primordial organisms are the following:

1. The protein must be present in both prokaryotes and eukaryotes.
2. The protein must be highly conserved across the eukaryotic–prokaryotic distance.
3. There must be intervening sequences in the eukaryotic version of the gene.
4. The protein must have been extensively studied structurally and biochemically, so that boundaries of its domains are known.

Several genes encoding metabolic enzymes have fulfilled these requirements. As already shown for GAPDH, these enzymes were present in the progenote and have been extremely conserved across the great evolutionary distance separating prokaryotes and eukaryotes. Also, as will be shown, several of the genes encoding these enzymes in eukaryotes have now been sequenced and found to contain intervening sequences. Perhaps most important, these enzymes were an early subject for X-ray crystallographers and enzymologists, and thus much is known about the structure and function of these proteins.

This latter feature of these ancient enzymes is of considerable value inasmuch as the rigor of a Blake-type analysis of intron–domain relationships is totally dependent upon the depth of understanding of the protein structures involved. High-resolution atomic coordinates have been available for several metabolic enzymes for a number of years (Adams et al., 1970; Hill et al., 1972; Branden et al., 1973; Buehner et al., 1973), and in fact, investigators have been able to use this protein data alone to discern evolutionary relationships among domains of these proteins. It should be pointed out that the understanding of protein structure required for such analyses depends almost as much on the *definitions* of certain levels of protein structure as it does upon the resolution of the protein data itself. To put it another way, the significance of a relationship between introns and domain borders is critically dependent on the definitions of those borders.

Some authors use the term *domain* loosely to refer to some region of a protein whose level of structure lies somewhere between secondary and tertiary (Schulz and Schirmer, 1979). Others demand that the function of the protein structure be known before it can be considered a domain (Banaszak et al, 1981). The following set of definitions is essentially that summarized in the excellent review of protein structure by Banaszak et al. (1981).

1. *Primary structure:* the amino acid sequence of a protein.
2. *Secondary structure:* local regions of stable regular folding. The α-helix, β-sheet, and hairpin loop are the most commonly recognized examples.
3. *Supersecondary structure:* two or more elements of secondary structure that form a recognizably recurring pattern. A unit consisting of an α-helix, β-sheet, and α-helix is an example.
4. *Domain:* one or more supersecondary structures that as a unit fulfill the additional requirements of (a) known function, (b) conformational *and* functional homology with other supersecondary structures, and (c) a structure consisting of a continuous segment of a polypeptide strand. (The peptide backbone cannot wander back and forth between domains.)

5. *Lobe:* one or more supersecondary structures that as a unit would likely retain its conformation even if the covalent linkage with the remainder of the protein were lost.
6. *Tertiary structure:* the three-dimensional structure of a protein molecule composed of a single polypeptide chain.
7. *Quaternary structure:* the three-dimensional relationship between subunits in an oligomeric protein.

In Blake's (1978) initial discussion of the possible relationship between exons and protein structure, he considered the possibility that exons might encode domains *or* supersecondary structures. If one is interested in Blake's idea from an evolutionary standpoint, the consideration of domains alone (as defined earlier) might be more appropriate. If a supersecondary unit is to confer some advantage when it is accidentally carried into a new parent protein (by unequal recombinations in the introns flanking the structure's exon), one might expect (1) the unit to have a useful function and (2) the structure of the unit to be independent enough of the parent protein that its function remains intact in its new environment. These latter attributes are simply those of domains.

How does one recognize a protein unit that is "independent enough" to be considered a domain? Banaszak and co-workers (1981) suggest that the peptide backbone must not "wander into another segment of the protein before returning to complete the domain structure." At the risk of overinterpreting this statement, it seems to suggest that the geometric course of the backbone is more important to the structural independence of a protein segment than the interactions of the amino acid residues. In some cases this is probably not true. If a polypeptide chain "wanders" into an adjacent domain for a few residues, without forming any hydrogen bonds or salt linkages, and then returns to the original domain to form 15 consecutive hydrogen bonds as part of a β-sheet, is the β-sheet structure significantly affected by the wandering residues? Conversely, if the backbone does not wander into another domain but its residues form 15 consecutive hydrogen bonds with a chain in an adjacent domain, can its structure be expected to remain intact if removed from the protein in which those hydrogen bonds are possible?

Often the geometry of the α-carbon chain and the likelihood of forming certain intrastrand interactions are closely related (as will be seen clearly in the following discussion of distance plots). However, we would argue that it is the number and strength of the bonds stabilizing a domain that is the important quantity in the domain's independence and that the α-carbon geometry per se is only as important as its ability to predict the former. Thus, we would amend the Banaszak domain criteria to require that a peptide strand be largely bonded to itself before being considered a domain.

Which bonds are important? Obviously, the ideal situation would be to evaluate the hydrogen bonds, electrostatic interactions, hydrophobic interactions, and so on, and obtain a precise understanding of the structural independence of a given protein segment. Although much progress is being made in this field, such an analysis is not currently possible. In any case, both the peptide backbone geometry and hydrogen bonding scheme are important in the folding of

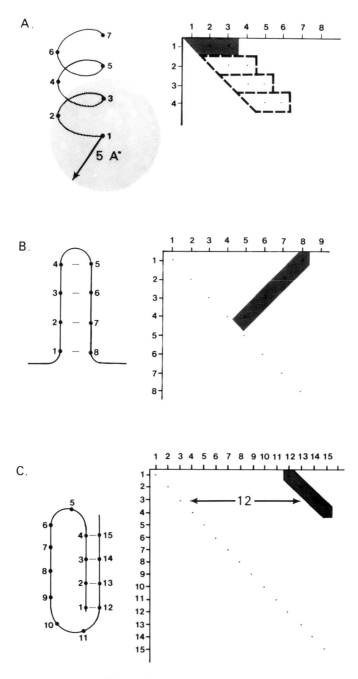

FIGURE 4.3. Appearance of basic elements of secondary structure on the α-carbon distance plot. α-Helix is represented as a band along the diagonal (A). Antiparallel β-sheet produces a bar of shading perpendicular to the diagonal (B). Parallel β-sheet yields a bar of shading parallel to the diagonal but displaced from it by the number of residues in the intervening loop (C).

protein segments to produce supersecondary structures. Machin and Phillips developed a simple way to display X-ray crystallography data that highlights backbone geometries consistent with secondary and supersecondary structures (Phillips, 1970). This "α-carbon distance plot" is constructed in the following way.

Positions along *both* the X and Y axes of the plot represent positions along the chain of α-carbon atoms of a protein's peptide backbone. For every α-carbon atom, the distance to every other α-carbon in the backbone is determined, and these distances are then represented schematically on the plot at the positions corresponding to each α-carbon pair. Consider how such a plot might reveal elements of secondary structure. Figure 4.3A shows the construction of a plot from a segment of α-helix. Starting at C1, the next three α-carbons are within 5 Å, but those of the next three (the next turn of the helix) are greater than 5 Å. If one indicates a distance of less than 5 Å by shading, this will produce a shaded area on the plot corresponding to the coordinates (1,1), (1,2), and (1,3). This situation will be identical for every α-carbon in the helix, producing a perfect band of shading along the portion of the diagonal corresponding to helix.

Now consider the situation when the peptide backbone turns back on itself to form an antiparallel β-sheet (Fig. 4.3B). Here the closest approach of α-carbons will be indicated at coordinates (4,5), (3,6), (2,7), and so on, which produces a band of shading *perpendicular* to the diagonal (and interesecting it at the point corresponding to the hairpin turn).

Figure 4.3C illustrates the final example, that of a peptide strand that has formed a parallel β-sheet. As in the case of the α-helix, the shaded area *parallels* the diagonal, but in this case is displaced horizontally by the number of residues in the loop preceding the sheet (12 in this example).

One may notice in the preceding examples that close interactions of α-carbons caused by various *secondary* structures result in densities on the distance plot *close to the diagonal*. This is simply because coordinates near the diagonal represent the interactions of α-carbons close to one another in the primary structure (the "local" folding referred to in the definition of secondary structure given earlier). Figure 4.4 demonstrates the rough generalization that as one moves away from the diagonal on a distance plot, the interactions reflect higher orders of structure. In Figure 4.4A, if the peptide strand is stretched out, the only near interactions would be along the diagonal. As one forms helices and sheets (Fig. 4.4B), the patterns previously explained are formed. If these combine into areas of supersecondary structure (Fig. 4.4C), large, roughly triangular areas appear along the baseline of the plot. Figure 4.4C also shows that as supersecondary units interact to form a tertiary structure, additional densities appear on the plot away from the diagonal (the rectangle in the upper right corner in this example).

Phillips (1970) recognized the distance plot as a simple way of representing the secondary structure and folding nuclei of proteins for which X-ray coordinates were available. Rossmann and Liljas (1974) noted that these plots were potentially valuable for detecting structurally homologous domains in different

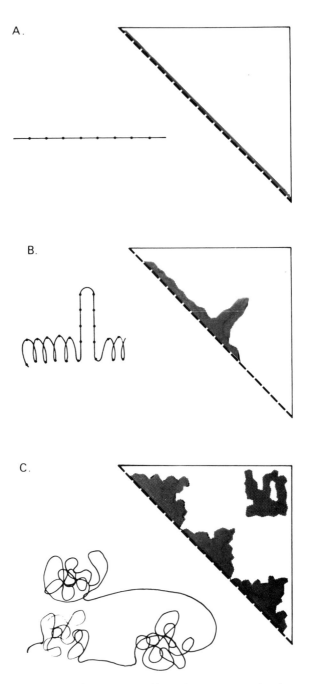

FIGURE 4.4. Appearance of increasing orders of structure on the distance plot. Primary structure (A) is represented by a diagonal line of points corresponding to each residue's relationship to itself. Secondary structure (B) produces patterns near the diagonal (see also Fig. 4.3). Supersecondary structure gives rise to triangles (composed of near and intermediate interactions) based upon the diagonal, whereas relationships between these supersecondary units (tertiary structure) yield a rectangle (of near and intermediate interactions) far from the diagonal (C).

proteins. They felt that domain borders and domain homology could "be readily recognized by eye where otherwise some sophisticated techniques might fail." Kuntz (1975) agreed that distance plots were useful for locating folding domains and went on to analyze the way in which interactions between secondary structures appear on such plots. He termed the patterns representing supersecondary structure *contour cells*.

Mitiko Go (1981) pointed out that distance plots were ideal for demonstrating the relationship between protein domains and intervening sequences suggested by Blake. When she analyzed a distance plot of β-globin, she found four protein units. The two intervening sequences in the β-globin gene mapped very close to two borders of these units. The suggestion was made that a third intron had been lost from the gene during evolution, a prediction that was confirmed when the leghemoglobin gene was sequenced and found to have an intron in the predicted position (Jensen et al., 1981). Shortly thereafter, a similar analysis (Go, 1983) of chicken egg-white lysozyme and its gene revealed a striking correlation between elements of supersecondary structure (termed *modules* by Go) and introns. The position of each of the three introns, when projected onto the protein, mapped very near a module–module border. This early experimental support of Blake's idea about the relationship of protein structure and gene structure established the distance plot (now often called the Go plot) as a powerful tool for demonstrating the location of structural domains within various proteins and the evolutionary relationships among them.

Figure 4.5 demonstrates the relationship between the distance plot patterns and structural features of the glycolytic enzyme lactate dehydrogenase. Panel A contains a stylized hydrogen-bonding scheme for the enzyme (Rossmann et al., 1971), and panel B consists of the corresponding distance plot (derived from the sequence data of Eventoff and Rossmann, 1975). In this plot the closest interactions of the peptide backbone (0–1.5 nm) are shown in gray; intermediate interactions (1.5–2.7 nm), in white; and distant interactions (>2.7 nm), in black. These parameters were chosen empirically to emphasize domain borders as well as some of the structural detail within domains. (The white lines on the plot are artifacts caused by the absence of coordinates for four atoms in the data base.) Panel C is a schematic representation of a few of the contour cells seen in the upper left corner of the distance plot. For any near interaction (gray area), one can determine the participating residues by simply measuring the *XY* coordinates of the area.

Such measurements reveal a thickening of the diagonal (feature 1 of panel C) to correspond to an α-helix (α-B in panel A). Similarly, the line parallel to the diagonal but displaced 15 residues to the right (feature 3 of panel C) reflects a parallel β-sheet interaction between two strands that are separated in the primary structure by 15 residues (β-A and β-C of panel A). Two lines somewhat perpendicular to the first two (features 2 and 4 of panel C) result from the interaction between the helix (α-B) as it curves back past the strands of β-sheet that it connects. The nearly perpendicular nature of these lines on the distance plot indicates that these are antiparallel interactions (helix going one direction, strands going the other; see also Fig. 4.3); the slight deviation from the true

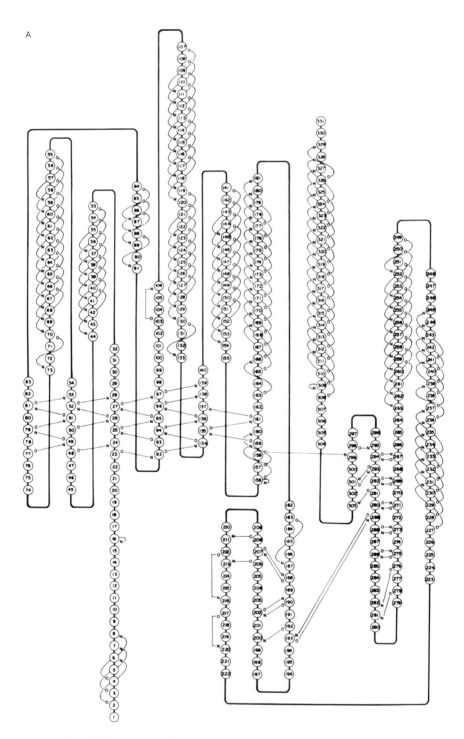

FIGURE 4.5. (A) The main chain hydrogen bonding scheme.

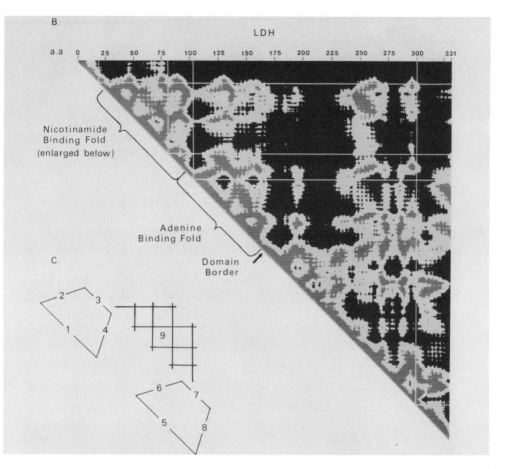

FIGURE 4.5. Relationship between the distance plot and main chain hydrogen bonding scheme of dogfish lactate dehydrogenase. The main chain hydrogen bonding scheme (A) emphasizes the areas of secondary structure within the LDH molecule. Positions in the primary structure are indicated by the residue numbers in the circles representing the individual amino acids. These areas of secondary structure can be found in the LDH distance plot (B) with the aid of the residue numbers at the top of the plot. Figure C schematically shows the distance plot appearance of nine structural features of the first mononucleotide binding fold of LDH (residues 1–100). (1) Helix B (residues 33–44); (2) interaction between β-sheet A (23–28) and α-helix B; (3) interaction between β-sheets A and B (48–53); (4) interaction between α-helix B and β-sheet B; (5) α-helix C (55–73); (6) interaction between sheet B and helix C; (7) interaction between sheets B and C (77–81); (8) interaction between helix C and sheet C; (9) interaction between helices B and C. (Panel A is reproduced with permission from Rossman et al., 1971. Panel B was constructed with data supplied by the protein Data Bank of the Brookhaven National Laboratory.)

perpendicular indicates an interaction between unlike secondary structures (helix and sheet; see also Kuntz, 1975).

The trapezoidal contour cell formed by elements 1–4 of panel C thus represents two strands of parallel β-sheet linked by 15 residues of α-helix. Further examination of panel A reveals an almost identical trapezoid just down the baseline from the first (features 5–8 in the schematic of panel C). This, not surprisingly, reflects an almost identical supersecondary unit of parallel β-sheet strands linked by α-helix. Feature 9 represents the parallel interaction of two helices. Taken together, features 1–9 represent a supersecondary unit of three β-sheets and two α-helices, which is, in fact, the nicotinamide binding region of the LDH molecule.

This nicotinamide binding region is just one of two mononucleotide binding folds in LDH. The other, which binds adenine, can be recognized on the Go plot as a pair of trapezoids just down the diagonal from the nicotinamide fold. The first trapezoidal contour cell of the adenine binding fold is larger than those of the nicotinamide fold because the α-helix (α-E, panel A) that links the first two β-sheet strands of the adenine fold (β-D and β-E) is longer. Taken together, the two mononucleotide-binding folds comprise the nicotinamide adenine dinucleotide (NAD) binding domain of LDH. This domain is seen on the distance plot as a rough triangle of gray and white interactions, approximately 165 residues on a side, based upon the diagonal. A similarly sized but less-well-defined triangle is based upon the lower half of the diagonal, and this group of interactions reflects the structure of the catalytic domain of LDH. The black area that separates these domains simply indicates that the residues within each domain are in general closer to each other (<27 Å) than to residues in the adjacent domain (>27 Å). The point at which the black area approaches the diagonal reflects the point at which the peptide backbone leaves the vicinity of one domain to enter that of another, in other words, the domain border.

In view of the structural homology between the two halves of the NAD binding domain, seen so readily with the aid of the distance plot, it is not surprising that some investigators (Eventoff and Rossmann, 1975; Harris and Waters, 1976) feel that the NAD binding domain evolved from a duplication of the mononucleotide binding fold.

Distance plots can reveal structural relationships *between* proteins as well. In Figure 4.6 the LDH Go plot has been repeated above the diagonal, and the plot of malate dehydrogenase (derived from the data of Birktoft and Banaszak, 1983) is given below. One can see instantly that the proteins are very similar structurally. MDH has an NAD binding domain making up the N-terminal half of the protein, and a very similar catalytic domain makes up the C-terminal half. However, the homology between the two proteins at the amino acid level is only 22–24 percent (Birktoft et al., 1982). This illustrates the advantage of distance plot comparisons over primary structure comparisons when searching for significant structural homologies.

Another Go plot comparison is shown in Figure 4.7. The same MDH plot is placed beside one for horse liver alcohol dehydrogenase (derived from the data of Branden et al. 1973). By sliding one plot past the other along the diagonal, one discovers a large degree of structural homology between the N-terminal

INTRON-DEPENDENT EVOLUTION OF PROGENOTIC ENZYMES 77

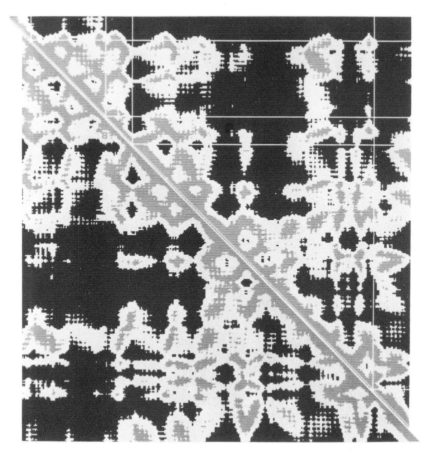

FIGURE 4.6. Distance plot comparison of dogfish lactate dehydrogenase and pig malate dehydrogenase. The α-carbon distance plot of MDH is given below the diagonal and that of LDH is above. The alignment maximizes the agreement of the structural features along the baseline. These plots were constructed with data supplied by the Protein Data Bank of the Brookhaven National Laboratory.

half of MDH and the C-terminal half of ADH. In fact, the NAD binding domain of the latter enzyme is found in the C-terminal portion. The structural homology of these NAD domains, coupled with the finding that the domain appeared to move *en bloc* from one end of the primary structure to the other during the evolution of these related proteins, led Rossmann et al. (1975) to suggest a gene fusion mechanism to explain the "sharing" of whole domains. Such a model required that an unequal recombination fall at the borders of the domains in the proper reading frame and seemed an unlikely event. Of course this was before the discovery of intervening sequences and RNA splicing, which could accomplish the same "domain sharing" with much greater frequency, providing that introns existed at the borders of the domains.

The structure of the ADH gene of maize has been determined (Dennis et al., 1984), and the positions of the nine intervening sequences (as projected onto

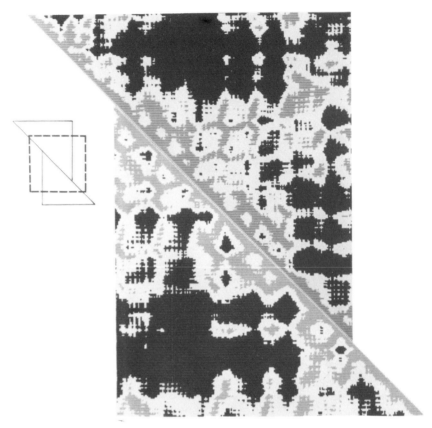

FIGURE 4.7. Distance plot comparison of pig malate dehydrogenase and horse liver alcohol dehydrogenase. The N-terminal portion of the MDH plot is given below the diagonal and the C-terminal portion of LADH is given above. The alignment maximizes agreement between the NAD binding domains of the two proteins. These plots were constructed with data supplied by the Protein Data Bank of the Brookhaven National Laboratory.

the protein structure) are also shown in Figure 4.8. One can see that an intron lies at the junction of the two mononucleotide binding folds as well as at the junction of the catalytic and NAD binding domains. It is easy to envision how an intron at the latter position could facilitate a "shuffle" of an intact NAD binding domain to another gene.

Figure 4.9 shows the distance plot of a related enzyme glyceraldehyde-3-phosphate dehydrogenase (derived from the data of Moras et al., 1975). Again, two large domains are readily apparent as white triangles based upon the diagonal. Examination of the contour cells within the first domain reveals a pattern reminiscent of the NAD binding domain of ADH, and it is thus not surprising that the N-terminal domain of GAPDH in fact binds NAD (Harris and Waters, 1976). The second domain in GAPDH is the catalytic unit (Harris and Waters, 1976). As before, the positions of the introns within the GAPDH gene (Stone et al., 1985b) are plotted on the distance plot. Again, an intervening sequence exists at the border of the NAD binding and catalytic domains as well as at the junction of the mononucleotide binding folds. Thus, it appears that the NAD binding domain, complete with the presumably ancient intron between the mononucleotide binding folds, was "shuffled" among these related genes during their evolution.

Branden (1986) has recently pointed out another noteworthy relationship between protein structure and gene structure in the cofactor binding domains of the group of dehydrogenases and kinases just discussed. As mentioned in the Go plot discussions previously, the basic structural theme of these domains is one of α-helix alternating with β-sheet. Specifically, enzymes with *di*nucleotide binding domains contain six recognizable strands of β-sheet connected by somewhat variable lengths of α-helix (and occasionally also additional small

FIGURE 4.8. Relationship of introns to structural features of alcohol dehydrogenase. The positions of the nine introns of the maize alcohol dehydrogenase gene are marked along the baseline of the distance plot of the horse liver ADH enzyme. Intron VII separates the two mononucleotide binding folds and intron IV separates the NAD binding domain from the catalytic domain. The distance plot was constructed with data supplied by the Protein Data Bank of the Brookhaven National Laboratory. (Intron positions are from Dennis et al., 1984.)

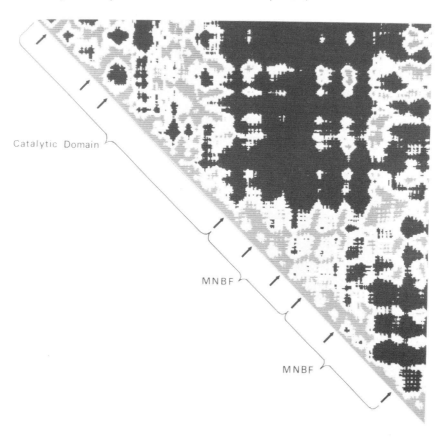

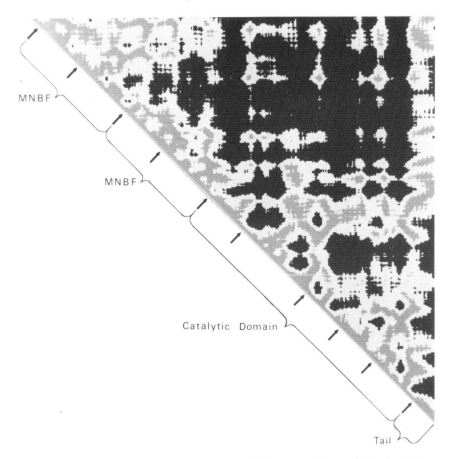

FIGURE 4.9. Relationship of introns to structural features of glyceraldehyde-3-phosphate dehyodrogenase. The positions of the 10 introns of the chicken GAPDH gene that interrupt the protein coding sequence are marked along the baseline of the distance plot of the lobster GAPDH enzyme. Intron IV separates the two mononucleotide binding folds, intron VI separates the NAD binding domain from the catalytic domain, and intron XI separates the helical tail from the remainder of the enzyme. Intron I interrupts the 5′ untranslated region of the gene and thus is not depicted in this figure. The distance plot was constructed with data supplied by the Protein Data Bank of the Brookhaven National Laboratory. (Intron positions are from Stone et al., 1985b.)

segments of β-sheet). Branden noticed that the introns of several of these genes often mapped to a homologous spot in the repeating α–β unit, namely, near the COOH end of the β-sheet strands. In addition, he noted that amino acids important to the function of these domains also often map near this spot. He proposes that these domains evolved from duplications of an intron-flanked segment of DNA encoding a single α–β unit and that splicing errors increased the probability of constructing important catalytic sites at the borders of these

INTRON-DEPENDENT EVOLUTION OF PROGENOTIC ENZYMES

units. Gilbert (1978) first pointed out how splicing errors could affect the evolution of a gene at the DNA level. His idea applied here would suggest that errors made during splicing of the mRNA from this mosaic gene caused a high mutation rate (at the protein level) near the borders of the α–β units. This would allow the ancient organism to try out a number of potentially useful residues in this border region without a mutation in the DNA. Any subsequent DNA mutation that increased the frequency of a favorable splicing "error" would itself be favorable and could thus fix a splicing error mutation in the DNA.

Branden's observation is depicted in Figure 4.10. The figure is derived from

FIGURE 4.10. Intron placement in the nucleotide binding domains of five progenotic enzymes. Shaded arrows represent the six β-sheets that are the most highly conserved features of the NAD binding domains. Thin lines indicate the lengths of peptide backbone joining the principal β-sheets. Intron positions of the five genes are plotted: GAPDH (o), ADH (x), LDH (■), PGK (▲), PK (●). (Redrawn after a figure from Branden, 1986.)

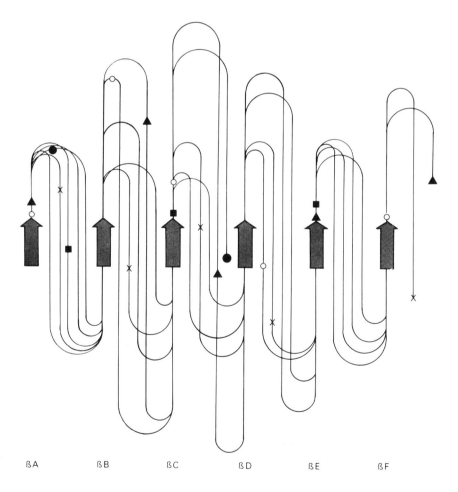

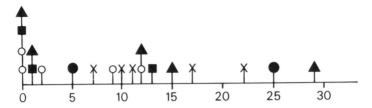

FIGURE 4.11. Relationship of introns to the carboxyl terminals of the principal β-sheets. The distance (in amino acids) from each intron to the carboxyl terminal of the preceding β-sheet is shown for five genes: GAPDH (o), ADH (x), LDH (■), PGK (▲), PK (●).

those in Branden's report but differs in the following ways. Intron data from five related genes are superimposed to reveal the relationships between intron placements in the various genes. Also, the actual lengths of peptide segments connecting the principal β-sheet strands are depicted to emphasize the variability of this structural feature in these related genes. Using the β-strands for alignment is justified by the data of Rossmann et al. (1975), which show the three-dimensional correspondence of the β-strands of the three dehydrogenases to be about three times greater than that for the interspersed α-helices.

Figure 4.10 reveals the greatest variation in distances between β-sheets to exist in the interval that links the two mononucleotide binding domains. The evolutionary significance of this is unclear, but it demonstrates how a relationship among introns of related genes might be missed if absolute distances from the beginning of the entire domain were compared. The figure has several other noteworthy features. First, the β-strands themselves are not interrupted by introns (as mapped from the gene to the protein). Second, for the six intervals following the β-strands, in the four proteins examined across the entire domain (24 intervals in all) there are 17 introns. This shows that the average exon length and average length of α–β unit are fairly compatible with the idea of a primordial α–β exon. The specific observation that introns are related to the COOH end of the principal β-sheets is shown in Figure 4.11. Plotting the distance from each intron to the COOH end of the preceding β-strand reveals a clustering of introns near the latter point. The average interval between the COOH ends of successive β-sheets in these genes is 27.7 (SEM = 2.13). Thus, if introns were randomly distributed in this interval we would expect their mean distance from the preceding β-strand to be half the interval, or 13.85 residues. The actual mean distance from β-strand to intron (as mapped to the protein) is 8.74 (SEM 1.9). The Student T test reveals this value to be significantly smaller than the "random" value at the $p \leq 0.01$ level.

To summarize, there are several pieces of data to suggest that introns facilitated the evolution of these progenotic genes.

1. Recognizable cofactor binding domains are shared by enzymes with differing catalytic domains.
2. These domains are often delimited by introns.

INTRON-DEPENDENT EVOLUTION OF PROGENOTIC ENZYMES

3. The homologous domains may be found at different points in the primary structure of the various enzymes.
4. The presumably duplicated halves of the dinucleotide binding domain are themselves delimited by introns.
5. The α–β units making up these mononucleotide binding folds are delimited by introns 17 out of 24 times, with a statistically significant relationship between the intron and the COOH end of the α–β unit.

Several investigators (Branden et al., 1985; Gilbert, 1985; Stone et al., 1985a) have taken this concordance of introns and recognizable structural features in these progenotic enzymes as strong evidence in support of Doolittle's idea that splicing existed before the divergence of prokaryotes and eukaryotes. Is this reasonable?

Could these introns be randomly distributed and the concordances mentioned earlier be figments of a biased imagination aided by a small sample size? After all, most arguments for concordance between intron placement and structural features invoke the possibility of loss and gain of introns, *after* the intron-dependent evolution of the genes, to explain the presence or absence of introns that do not fit the model at hand. For example, we noted three intron–protein relationships in our report (Stone et al., 1985a) on the structure of the GAPDH gene. What about the other seven introns that interrupt the protein coding sequence? With 10 introns scattered about, isn't it fairly likely that three of them will fall "near" domain borders to form a "striking concordance"?

Consider a 330-residue protein with four recognizable structural units. What is the likelihood that three of 10 introns will fall within three zones, each 15 amino acids wide, separating the four structural units? Figure 4.12 shows the probability of this event (which is essentially that for GAPDH) to be 6.8 percent. LADH also has three zones in which introns would be expected: two to delimit the NAD domain and one to join the MNBFs. The probability of finding introns in these locations is a similar 6.8 percent, but the probability of these introns occurring in *both* of these genes is only 0.5 percent. Thus, as more genes from this related group are sequenced, the probability that the observed intron patterns are due to chance may become very small indeed.

FIGURE 4.12. Probability of three of 10 introns falling in three 15-aa zones in a 330-aa protein. The probability that one intron will fall in a shaded zone is 45/330. The probabilities of a second and third intron falling in the remaining zones are 30/330 and 15/330, respectively. The probability of all three events is (45/330) (30/330) (15/330) = 5.6×10^{-4}. However, since there are 10!/(7!)(3!) = 120 ways to select three introns from 10, the probability of three of the 10 introns falling into the three shaded zones is $120 \times 5.6 \times 10^{-4}$ = 6.8 percent.

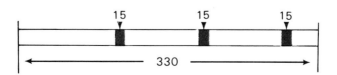

Table 4.1. Exon Sizes in Five Progenotic Genes

Exon	GAPDH	ADH	LDH	PGK	PK
1	39	133	unk	unk	unk
2	41	137	118	51	170
3	100	47	174	156	92
4	108	326	176	145	132
5	90	83	118	104	187
6	116	76	124	120	271
7	82	62	652	115	151
8	152	96		180	153
9	79	162		178	167
10	84	481		99	182
11	98			475	795
12	297				

Another possibility is that introns are not randomly dispersed but that their pattern reflects some constraint other than the structure of the protein encoded by their gene. For example, Naora and Deacon (1980) analyzed exon lengths of approximately 80 genes, "mostly from higher eukaryotes," and found these exons to fall into discrete size classes. The three major classes were 52 ± 2, 140 ± 2, and 223 ± 3. They noted that these sizes corresponded roughly to the sizes of the nucleosome linker region, core particle, and entire nucleosome, respectively. In view of the data, one might imagine some predisposition to intron insertion based upon chromatin structure, and if a repeating structural feature of the gene product (such as an $\alpha-\beta$ unit) accidentally had the same period, one might mistake an actual relationship to the chromatin for an apparent relationship to the gene product.

Unfortunately, as noted by Blake (1985), Naora and Deacon included multiple copies of some genes as well as genes with multiple repeats of the same exon, both of which would tend to bias the data. When Blake repeated the analysis using 20 genes from different protein families and excluding genes with highly repetitive exons (collagen and vitellogenin), the result was a unimodal distribution of exon sizes centered on 135 base pairs.

In this context, it is interesting to examine the distribution of exon sizes for the five genes discussed in this section. Table 4.1 gives the data summarized graphically in Figure 4.13. The 5' end of three of the genes has not been determined and thus the lengths of the first exon (or exons) is unknown. Also, the maize ADH gene lacks a poly(A) addition signal; as a result, the length of the last ADH exon in the table represents a minimum length. Examination of the 48 exon lengths given in Table 4.1 reveals only one that falls into one of Naora and Deacon's "major size groups," while the mean length of the protein coding exons is 127.3, very near the 135 found by Blake. Another noteworthy feature of these genes is the size of the 3' exons. In every case, the last exon is the longest, and the average length of the last exons is 4.25 times the average length of the remaining exons.

These terminal exons are intriguing in several respects. Much of the length of these exons is contributed by a substantial 3' noncoding sequence. In some genes (Yaffe et al., 1985) such noncoding sequences are highly conserved and are thought to harbor information used to direct posttranscriptional regulation of gene expression. In this context, it is interesting that no intron exists between the coding and noncoding sequences at the 3' end of the gene. That is, it does not appear that a 3' regulatory element was added to an existing progenotic gene by the same unequal-crossover mechanism proposed for the gene itself (which would have left an intron at the interface of the gene and the regulatory element). Instead, it seems that some 3' noncoding sequences were essential for expression of the earliest version of a gene. When these early genes were joined by unequal recombination, the downstrean portion of the new gene would retain its 3' noncoding sequences (without a separating intron), whereas the 3' noncoding sequences of the upstream portion of the new gene would become part of the intron resulting from the unequal recombination.

FIGURE 4.13. Disparity between the length of the 3' terminal exons and all other exons of five progenotic genes. Mean of all exons of each gene is given by a white bar. The mean of all except the last exon of each gene is given by a shaded bar. The length of the 3' terminal exon is marked (ဟ) above the shaded bars. The bars labeled "all" give the averaged data from all five genes.

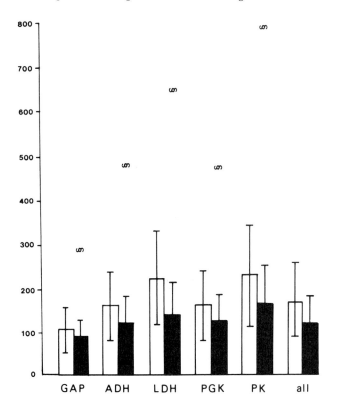

Of course, some genes *do* have an intron at the junction of the 3' noncoding region and the coding sequences (Jones et al., 1985), and a loss of an intron from this position of the five genes discussed earlier cannot be ruled out.

Another noteworthy consequence of the large 3' exons of these progenotic genes is that they constitute a strong argument against random insertion as a mechanism for the origin of introns. That is, with a mean exon length of 170 nucleotides, one would not expect a random insertion mechanism to spare several hundred nucleotides of noncoding sequence in these five genes.

As already mentioned, a troubling problem with using protein–gene concordance to argue for intron-mediated evolution is the apparent loss, gain, and movement of introns. For example, GAPDH and ADH have their NAD binding domain gene segments crisply delimited from their catalytic domain gene segments by introns (Branden et al., 1984; Stone et al., 1985b). Thus, if introns functioned in the sharing of the domains among these genes, one would expect the mouse LDH gene to have an intron between the gene segments encoding its NAD binding and catalytic domains. Unfortunately, no intron is present in this position in the mouse gene (Li et al., 1985). One is tempted to explain this absence as an evolutionary loss of the intron, and indeed this explanation may be supported when the genes for the other two LDH isozymes are sequenced. Similarly, when one compares the two introns of a nematode GAPDH gene (Yarbrough et al., 1987) with those of the chicken gene (Stone et al., 1985b), one finds a perfect match with one intron and a disturbing mismatch with the other. Specifically, the second nematode intron and the eleventh chicken intron split a homologous tryptophan codon at exactly the same point. In contrast, the first nematode intron interrupts the coding sequence 20 nucleotides upstream from the position of the seventh chicken intron and 62 nucleotides downstream from the sixth chicken intron.

At least four explanations can be advanced for the mismatch:

1. Chicken intron 7 is ancient but the nematode lost it and all but one of the other ancient introns and gained its intron 1 relatively recently (after the divergence of the chicken and nematode).
2. Nematode intron 1 is ancient but the chicken lost it and gained its intron 7 recently.
3. Both introns are recent.
4. Both introns are derived from the same ancient intron but their positions were displaced by sliding splice junctions as proposed by Craik et al. (1983).

As discussed by Doolittle in Chapter 3 of this volume, it is hazardous to argue for an intron's gain or loss on the strength of a few sequences. However, as more and more gene sequences are obtained from organisms scattered about the phylogenetic tree, the ontogeny of introns should become increasingly apparent. And although the evidence at this point in our understanding seems to favor intron loss over intron gain as the explanation for differences in intron patterns among related genes, we agree with Blake (1985) that intron gain has not been completely ruled out.

As discussed elsewhere (Stone et al., 1985a), we like to classify introns for the purposes of evolutionary discussion into two broad groups: (1) associative introns (type A) that came into being as two previously distant gene segments were juxtaposed by an unequal crossover and (2) divisive introns (type B) that arose when DNA insertions divided previously contiguous gene segments but were tolerated because the organism managed to splice them out at the RNA level. We further subdivide the type A introns based on the types of protein structures that were associated by the recombination event that gave rise to the intron. Type A1 introns join gene segments that encode proteins whose structure in the total protein is dependent upon interactions with other peptide segments. The introns that link the $\alpha-\beta$ blocks within a mononucleotide binding fold are examples of A1 introns. Type A2 introns join gene segments that encode homologous domains or supersecondary units that are relatively independent structurally and that also appear to result from a duplicative unequal recombination. Such A2 introns can be found linking the three domains of ovomucoid (Stein et al., 1980), the four domains of the immunoglobulin heavy chain (Barker et al., 1980), the numerous 18 amino acid domains of collagen (Ohkubo et al., 1980), and the three domains of α-fetoprotein (Tilghman, 1982). Type A3 introns also link gene segments encoding structurally independent peptides, but here the domains or supersecondary units are nonhomologous and serve to associate domains of different function to create a higher-order function. The introns linking the NAD binding domains with the catalytic domains of dehydrogenases are examples.

In closing this chapter, we would like to consider some features of the biology of organisms that preceded the progenote. In the foregoing pages, we have concentrated on a mechanism for constructing and then sharing domains to form several related genes in an early organism. Many genes of the type we discussed are essential for the simplest metabolic processes we associate with life. What mechanism was producing the energy to drive the reactions of an organism during the initial assembly of the glycolytic enzyme genes? If an energy-producing system existed that was independent of carbohydrate oxidation, why would the first steps toward the present systems have been advantageous?

We would like to suggest a mechanism whereby functional energy-producing enzymes could have existed *before* their genes. Blake (1985) and Lonberg and Gilbert (1985) have suggested that the small primordial exons may have been translated into small oligopeptides. We would like to extend this idea to propose that such oligopeptides may have aggregated in the cytosol to form a functional enzyme, much as the numerous-component ribosomal proteins aggregate to form a functioning ribosome.

The basis for the idea lies in the previous discussion on protein structure, specifically, in the distinction between an element of tertiary structure, the lobe, and quaternary structure. Banaszak et al. (1981) use the term *lobe* to describe a "larger subassembly of the polypeptide chain that appears to be stable even if the chain connections to other parts of the protein were to be cleaved . . . [and which] has all the structural characteristics of an entire globular protein with respect to distribution of types of amino acids." If one were simply to break the peptide bond at the border of such a lobe, the result would be a dimeric enzyme composed of different subunits. Inasmuch as the single peptide bond would provide only a fraction of the force holding the two lobe–subunits together, one might expect that the subunits would continue to associate as before and would continue to function. Moreover, since their structures are so independent, the lobes probably fold using separate nucleation centers, and if denatured, one might not be surprised if the dimeric structure reformed despite the loss of covalent linkage between the two structures.

Now consider a hypothetical primitive organism whose DNA encodes two such lobes on separate genes. One such gene might encode an NAD binding protein, and another a glyceraldehyde-3-phosphate binding protein. These two proteins, separately translated, could associate in the cytoplasm of the organism to form a primitive glyceraldehyde-3-phosphate dehydrogenase, without an actual GAPDH gene existing in the DNA. To carry the idea further, the gene encoding the substrate binding domain could be duplicated and subsequently mutated to produce several different "catalytic" domain genes. The products of the latter could each associate (posttranslationally) with the NAD binding protein to form a family of functioning dehydrogenases in the organism without any complete dehydrogenase genes in the DNA.

Some of these heterodimeric enzymes would be very useful to the organism and some would not. If an unequal recombination occurred that juxtaposed the two genes encoding the subunits of a useful dehydrogenase, they could then be transcribed on a single piece of RNA (linked by an A3 intron that would be removed by splicing), allowing the enzyme to be translated as a single polypeptide. This mutation would free the favorable catalytic domain from competition with other domains and increase the ratio of the favorable dehydrogenase to unfavorable ones in the cytoplasm. It would also produce a single mosaic gene in the DNA encoding an enzyme that had already been functioning in the organism for many generations.

Just as an enzyme could exist before its gene via this mechanism, we believe certain domains could have existed in the cytoplasm of primitive organisms without being contiguously encoded in the DNA. That is, we can imagine a piece of DNA encoding an α-helix–β-sheet subunit in a primitive organism. Once translated, three of these peptides might associate and have some affinity for ATP (Fig. 4.14); six might associate and have affinity for NAD. Any number of these aggregates could exist in the cytosol of the organism simultaneously and be functioning as crude protein domains. Formation of the more useful aggregates would be favored by unequal crossovers that juxtaposed the genes encoding the subunits of the aggregate, allowing them to be transcribed on a single message (linked by A1 introns), spliced, and finally translated as a single polypeptide.

The mechanism we propose here is actually open to experiment. That is, after constructing a suitable enzyme-deficient host bacterium, one could transform this host with plasmids bearing various fragments of an enzyme gene and thereby obtain the protein products of these fragments. Thus, one could demonstrate whether independently translated NAD and catalytic domains will ac-

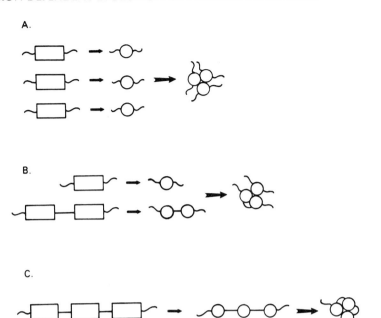

FIGURE 4.14. Proposed mechanism whereby a functional enzyme could antedate its gene. (A) Small exons are independently translated into protein subunits that aggregate in solution to form a functional multimeric enzyme. (B) Unequal recombination unites two of these exons in a single transcription unit, improving the likelihood that the functional aggregate will form. (C) A second unequal crossover brings the third exon into the transcription unit and completes the gene, long after the functional enzyme appeared in the organism.

tually associate in solution to form a functioning enzyme. If the latter could be demonstrated, one could go on to test whether the products of individual exons would form crudely functional aggregates.

Acknowledgments

The authors thank Drs. Charles Lawrence and Jean Pierre Bretaudiere for their assistance in constructing the α-carbon distance plots, and Jan Case for her excellent illustrations.

References

Adams, M. J., Ford, G. C., Koekoek, R., Lentz, P. J., McPherson, A., Rossmann, M. G., Smiley, I. E., Schevitz, R. W., and Wonacott, A. J. (1970). Nature 227:1098.

Banaszak, L. J., Birktoft, J. J., and Barry, C. D. (1981). In Protein–Protein Interactions. Frieden, C., and Nichol, L. W., eds. Wiley, New York, p. 31.

Barker, W., Ketcham, L., and Dayhoff, M. (1980). J. Molec. Evol. 15:113.
Birktoft, J. J., and Banaszak, L. J. (1983). J. Biol. Chem. 258:472.
Birktoft, J. J., Fernley, R. T., Bradshaw, R. A., and Banaszak, L. J. (1982). Proc. Natl. Acad. Sci. USA 79:6166.
Blake, C. C. F. (1978). Nature 273:267.
Blake, C. C. F. (1985). Int. Rev. Cytol. 93:149.
Branden, C. I., (1986). In Current Communications in Molecular Biology: Computer Graphics and Molecular Modeling, Fletterick, R., and Zoller, M., eds., Cold Spring Harbor Laboratories, Cold Spring Harbor, N.Y., p. 45.
Branden, C. I., Eklund, H., Nordstrom, B., Boiwe, T., Soderlund, G., Zeppezauer, E., Ohlsson, I., and Akeson, A. (1973). Proc. Natl. Acad. Sci. USA 70:2439.
Branden, C. I., Eklund, H., Cambillau, C., and Pryor, A. J. (1984). EMBO J. 3:1307.
Buehner, M., Ford, G. C., Moras, D., Olsen, K. W., and Rossman, M. G. (1973). Proc. Natl. Acad. Sci. USA 70:3052.
Craik, C. S., Rutter, W. J., and Fletterick, R. (1983). Science 220:1125.
Dennis, E. S., Gerlach, W. L., Pryor, A. J., Bennetzen, J. L., Inglis, A., Llewellyn, D., Sachs, M. M., Ferl, R. J., and Peacock, W. J. (1984). Nucleic Acids Res. 12:3983.
Doolittle, W. F. (1978). Nature 272:581.
Eventoff, W. and Rossmann, M. G. (1975). In CRC Critical Reviews of Biochemistry, Vol. 3. Fasman, G. D., ed. CRC Press, Cleveland, p. 111.
Gilbert, W. (1978). Nature 271:501.
Gilbert, W. (1985). Science 228:823.
Go, M. (1981). Nature 291:90.
Go, M. (1983). Proc. Natl. Acad. Sci. USA 80:1964.
Harris, J., and Waters, M. (1976). In The Enzymes, Vol. 13. Boyer, P., ed. Academic Press, New York, p. 1.
Hill, E., Tsernoglou, D., Webb, L., and Banaszak, L. J. (1972). J. Molec. Biol. 72:577.
Jensen, E. O., Paludan, K., Hyldig-Nielsen, J. J., Jorgensen, P., and Marker, K. A. (1981). Nature 291:677.
Kuntz, I. D. (1975). J. Am. Chem. Soc. 97:4362.
Li, S. S. L., Tiano, H. F., Fukasawa, K. M., Yagi, K., Shimizu, M., Sharief, F. S., Nakashima, Y., and Pan, Y. E. (1985). Eur. J. Biochem. 149:215.
Lonberg, N., and Gilbert, W. (1985). Cell 40:81.
Margulis, L. (1970). Origin of Eucaryotic Cells. Yale University Press, New Haven.
Moras, D., Olsen, K., Sabeson, M., Ford G., and Rossmann, M. G., (1975). J. Biol. Chem., 250:9137.
Naora, H., and Deacon, N. J. (1982). Proc. Natl. Acad. Sci. USA 79:6196.
Ohkubo, H., Vogeli, G., Mudryj, M., Avvedimento, E., Sullivan, M., Pastan, I., De Crombrugghe, B., (1980). Proc. Natl. Acad. Sci. USA 77:7059.
Phillips, D. C. (1970). British Biochemistry Past and Present. Goodwin, T. W. ed. Academic Press, New York, p. 11.
Rossmann, M. G., Adams, M. J., Buehner, M., Ford, G. C., Hackert, M. L., Lentz, P. J., McPherson, A., Schevitz, R. W., and Smiley, I. E., (1971). Cold Spring Harbor Symp. Quant. Biol. 36:179.
Rossmann, M. G., and Liljas, A., (1974). Ann. Rev. Biochem. 43:475.
Rossmann, M. G., Liljas, A., Branden, C., and Banaszak, L. (1975). In The Enzymes, Vol. 11. Boyer, P., ed. Academic Press, New York, p. 61.
Rossmann, M. G., Moras, D., and Olsen, K. W. (1974). Nature 250:194.
Schulz, G. E., and Schirmer, R. H. (1979). Principles of Protein Structure. Springer Verlag, New York.

Stein, J. P., Caterall, J. F., Kristo, P., Means, A. R., and O'Malley, B. W. (1980). Cell 21: 681.
Stone, E. M., Rothblum, K. N., Alevy, M. C., Kuo, T. M., and Schwartz, R. J. (1985b) Proc. Natl. Acad. Sci. USA 82:1628.
Stone, E. M., Rothblum, K. N., and Schwartz, R. J. (1985a) Nature 313:498.
Sudhof, T. C., Goldstein, J. L., Brown, M. S., and Russell, D. W. (1985). Science 228:815.
Tilghman, S. M. (1982) 27th Scient. Rep. Inst. Cancer Res. 125.
Woese, C. R., and Fox, G. E. (1977). J. Molec. Evol. 10:1.
Yarborough, P. O., Hayden, M. A., Dunn, L. A., Vermersch, P. S., Klass, M. R., and Hecht, R. M. (1987). Biochim. Biophys. Acta 908:21.

5

Intervening Sequences in Molecular Evolution

JOSEPH P. STEIN, MAXWELL J. SCOTT, AND BERT W. O'MALLEY

THE DISCOVERY OF SPLIT GENES

Like 1953, the year 1977 will be looked upon as a landmark year for molecular biology by those who study the history of science. Until that year the possibility that eukaryotic genes might be noncontiguous had probably never occurred to molecular biologists. Numerous bacterial genes had been isolated and sequenced and the colinearity of chromosomal coding information with the primary amino acid sequence of proteins had been firmly established. Indeed, a few animal cell messenger RNAs had been sequenced, in whole or in part, and like bacterial mRNAs, were found to consist of an uninterrupted protein-coding sequence defined by initiation and termination codons, flanked at either end by short, nontranslated sequences. So there seemed to be no reason to expect gene organization in eukaryotes to be any different from prokaryotes.

However, a slightly unsettling report was published early in 1977 that demonstrated that some of the 28S rRNA genes of *Drosophila* contained 5.4 kb of DNA of unknown function that was not found in the 28S rRNA molecule itself (Glover and Hogness, 1977). Probably because there was no evidence that these unusually large 28S rRNA genes were even expressed, little significance was attached to this report by the biochemical community. However, one began to suspect that the organization of genetic information in higher organisms might not be as orderly as originally supposed when several groups reported in August and September of that year that several different viral mRNAs contained 5' terminal sequences that were complementary to sequences in the viral genome remote from the major coding sequence of the viral mRNAs. For example, Berget et al. (1977) used purified late hexon mRNA from adenovirus-infected cells in R-looping experiments to demonstrate that a 160-bp 5' end "tail" hybridized to three short segments of the R strand of adenoviral DNA located upstream of the hexon gene (at 16.8, 19.8, and 26.9 map units). They correctly surmised that these short segments were probably spliced to the body of the mRNA during posttranscriptional processing. Furthermore, "a plausible model for the synthesis of the mature hexon mRNA would be the intramolecular joining of these short segments to the body of the hexon mRNA during the processing of a nuclear precursor to generate the mature mRNA." Aloni et al. (1977) likewise showed that late SV40 mRNAs also contained a similar arrangement of leader sequences that were noncontiguous with the coding sequences. Still, these were viral mRNAs and there remained the quite likely possibility that these "spliced" leader sequences were a peculiarity of the 5' end of these mammalian viral mRNAs.

Before molecular geneticists had much time to reflect upon the universality of these findings, the next six-month period resulted in the independent publication of the "split gene" concept by seven laboratories throughout the world. In November, 1977, both Chambon (Breathnach et al., 1977) and Carey (Doel et al., 1977) published their findings that the chicken ovalbumin gene contained at least two intragenic DNA inserts of noncoding DNA sequences! There rapidly followed two more such reports in December, 1977. Tonegawa (Brack and Tonegawa, 1977) found that a rearranged mouse lambda chain immunoglobin gene contained a 1,250-bp insert, and Flavell (Jeffreys and Flavell, 1977) showed that the 438-bp β-globin structural gene in rabbits is interrupted by a 600-bp insert of nonrepetitive, nonglobin DNA, and that this β-globin gene structure was found in all rabbit tissues. In the ensuing several months, Axel (Weinstock et al., 1978) and O'Malley (Dugaiczyk et al., 1978) confirmed and extended these findings to show that the hen ovalbumin gene coding sequences were in fact broken up by multiple intervening sequences. The latter group demonstrated that the "pieces" of coding sequences were spread out over a large region of the genome that represented the primary transcription unit.

Looking back, it is hard now to recall the quantum leap of enlightenment that these reports precipitated, precisely because our knowledge of eukaryotic gene structure and regulation has since expanded so much in so few years. The concept of a mammalian gene with noncontiguous coding information was so novel in 1977 that early Southern blots of hen genomic DNA with an ovalbumin cDNA probe, which resulted in a greater number of hybridizing fragments than expected, were quite disturbing to investigators in our laboratory. (It should be noted that these experiments were performed prior to the publication of the reports of spliced viral mRNAs). Once the possibility of multiple ovalbumin gene copies had been eliminated, we still thought it likely that the multiple hybridizing fragments were due to some spurious hybridization reaction. By the time these experiments had been repeated numerous times and the several investigators involved were finally convinced that restriction endonuclease recognition sites existed in the hen ovalbumin gene that were not present in its cDNA, personal communications and meeting presentations began to indicate that other investigators were observing the same phenomena. Even when we had exhaustively mapped the hen ovalbumin gene and determined that there were in fact seven intragenic insertions of noncoding DNA sequences, all the ovalbumin gene fragments observed on Southern digests were still unexplained. For example, a 1.8-kb EcoRl fragment of the ovalbumin gene, when used as a hybridization probe, was found to hybridize to a 1.3-kb genomic EcoRl fragment as well as itself. It was finally shown that this 1.3-kb fragment derived from a variant ovalbumin gene copy, present at a much lower frequency in the chicken genome, that had an additional EcoRl cutting site located in intervening sequence E (Lai et al., 1979a). This finding confirmed the existence of a new type of genetic allele that had no phenotypic representation, as the mutation occurred only in the newly discovered intervening sequences of genes rather than the peptide coding region. Such an allele could not be demonstrated by analysis of the sequences of the mRNAs, cDNAs, or proteins, but only by a detailed analysis of the cloned intervening sequences of the native gene. The analysis of such allelic differences in intervening sequences has since expanded and blossomed into an area of study itself. Numerous laboratories now study restriction fragment length polymorphisms (RFLPs), particularly as markers for locating and analyzing genes responsible for human diseases.

EARLY HYPOTHESES REGARDING INTERVENING SEQUENCES

Since the existence of these split genes in eukaryotes, now well ingrained in molecular genetics, was then perceived as being extremely novel, much speculation immediately centered on the origin and function of introns. Several early hypotheses purported to explain the existence of intervening sequences. Gilbert, who coined the term *intron* for such sequences, proposed that their actual function was in essence to speed evolution (Gilbert, 1978). Introns could accomplish this in two ways. First, in certain regions of the introns, single base changes could dramatically alter the splicing pattern of the transcript, resulting in the deletion or addition of whole segments of polypeptide. Thus, during the course of evolution, relatively rare single base mutations could generate novel proteins much more rapidly than would be possible if no splicing occurred. Second, since the presence of introns considerably lengthens the distance of any two exons within the same gene from each other, genetic recombination of exons should be enhanced. If exons are sequences that code for protein domains of particular function, then recombination within introns would tend to sort these functions independently, leading to a more rapid generation of protein products with novel combinations of functional groups. Implicit in this view was the idea that introns were the means (and the remnants) whereby functional genes were assembled from pieces of DNA that separately coded for some structural component necessary for the function of the final protein product. Thus, introns would be considered as frozen remnants of evolutionary history (static viewpoint) or as sites of continuing evolution (dynamic viewpoint). A related but somewhat more restrictive view was espoused by C. C. F. Blake, who proposed that exons corresponded to integrally folded protein units called domains or supersecondary structures (Blake, 1978).

Another early hypothesis was that introns were "mutational sinks" that provided a bulk of nonessential DNA that could accumulate mutations without harming the function of many gene products. Perhaps this idea was best explained by W. Ford Doolittle, who proposed that the split organization of eukaryotic genes was present before eukaryotic and prokaryotic genomes diverged about 2.5–3 billion years ago (Doolittle, 1978). The role of introns was to ensure that the transcripts of exons, which were (hypothetically) reiterated but often incorrectly replicated and transcribed, would be at least occasionally assembled so as to template functional proteins. This hypothesis requires the assumption that replication, transcription, and translation would have been relatively unfaithful in the last common ancestor of prokaryotes and eukaryotes, which might then have allowed rapid evolution. As these three enzymatic processes became more faithful, such insurance became less necessary and the replication and transcription of redundant and noninformational DNA became irrelevant and burdensome. Consequently, there was pressure to eliminate these DNAs, with the net result that introns should be decreasing in length and number with time. It appears that some evidence, based on comparative structural analysis of genes, can be found to support each of these hypotheses partially.

THE OVOMUCOID GENE AND DOMAIN STRUCTURE

Although it is not possible to examine the function of intervening sequences by experimentation in the classical sense, we were among the first investigators to examine critically some of these hypotheses of intron function using DNA sequence and protein structural data. For years this laboratory had focused on the hen oviduct as a model system for studying the structural organization of functionally related genes in a single tissue. Indeed, we cloned the hen ovomucoid gene shortly after the ovalbumin gene, and its structure proved to be very similar (Lai et al., 1979b). In fact, despite the fact that ovomucoid is functionally unrelated to ovalbumin and less than half the size of ovalbumin, the ovomucoid gene contained seven intervening sequences interspersed among eight exons, just like ovalbumin. During the latter stages of our sequence analysis of the cloned ovomucoid gene, the sequence of the secreted chicken ovomucoid protein was published (Kato et al., 1978). This permitted us to analyze the structure of the ovomucoid gene in light of the protein sequence and resulted in a unique opportunity to examine the *raison d'être* of intervening sequences. We will summarize this analysis in the next few paragraphs, as well as recent data on a related chicken gene and their implication for our arguments concerning the evolution of the ovomucoid gene.

Mature ovomucoid mRNA consists of 821 nucleotides (Caterall et al., 1980). A 53-nucleotide 5' leader sequence occurs prior to the AUG initiator codon, followed by a 72-nucleotide signal peptide sequence that is translated but re-

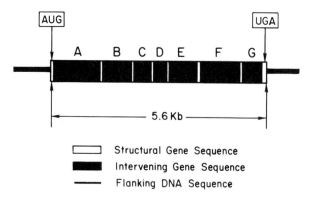

FIGURE 5.1. Structural organization of the chicken ovomucoid gene. Ovomucoid exon sequences are shown as *open bars;* the intron sequences, as *solid bars;* and the flanking genomic DNA, as a *solid line.* The seven introns are labeled A through G. Overall length of the ovomucoid gene is 5.6 kb.

moved from the secreted polypeptide during the secretion process. The 558 nucleotides that code for the secreted ovomucoid are next, followed by 138 nucleotides of 3′ noncoding sequence. The gene encoding this mRNA is 5.6 kb in length, and its structural organization is depicted in Figure 5.1. There are seven intervening sequences interspersed among the eight exons. These exons vary widely in length; the largest, exon 8, is nine times longer than the shortest, exon 2, which is notable because it is only 20 nucleotides in length. Remarkably, only 15 percent of the ovomucoid gene sequence is represented in the mature mRNA.

Avian ovomucoids are a family of glycoproteins that account for about 10 percent of the protein content of all bird egg whites (Feeney and Allison, 1969; Feeney, 1971). Chicken ovomucoid has been shown to be responsible for most of the trypsin inhibitory activity of chicken egg white (Lineweaver and Murray, 1947). For several years, Laskowski and co-workers had been investigating the comparative biochemistry of avian ovomucoids, and while we were completing the sequence analysis of the ovomucoid gene, they determined the amino acid sequence of the secreted chicken ovomucoid (Kato et al., 1978). These investigators noted a general sequence homology among ovomucoids and several mammalian submandibular and seminal plasma trypsin inhibitors. Specifically, the sequence of chicken ovomucoid, shown in Figure 5.2, was divided into three structural domains. Each domain is capable of binding one molecule of trypsin or some other serine protease, so the correlation of functional and structural domains is good. Whereas each domain contains three disulfide bonds in almost identical positions, the sequence homology between domains I and II is far stronger than that between domains I and III or II and III. Laskowski used these observations to propose that distinct intragenic doubling events of a primordial gene resulted in the present ovomucoid gene. We were thus in a position to examine this hypothesis more critically by comparing our ovomucoid genomic DNA sequence with the amino acid sequence of the secreted polypeptide, and also to assess the hypotheses regarding intron function with respect to this specific gene (Stein et al., 1980).

The illustration of the amino acid sequence in Figure 5.2 is depicted so that the functional, trypsin-binding domains are clearly delineated. They are depicted (in two dimensions) as globular regions held together by disulfide bonds and separated by short, connecting peptides. An analysis of the positions of the intervening sequences within the ovomucoid gene determined that several introns interrupted the portions of the gene sequence that code for the connecting peptides. We used the locations of the introns to redefine the three functional domains of chicken ovomucoid. These domains, depicted separately in Figure 5.3, are aligned to show the structural similarities as well as the similar posi-

FIGURE 5.2. Amino acid sequence of the secreted chicken ovomucoid. The single-letter amino acid code of Dayhoff is used. The drawing is meant to depict three similar structural regions held together by disulfide bonds *(solid lines)* separated by short, connecting peptides.

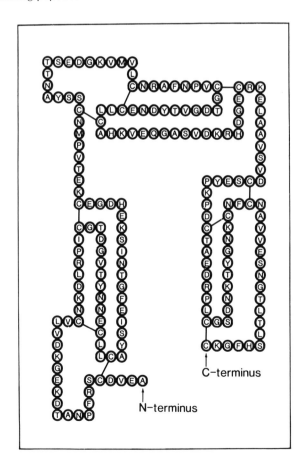

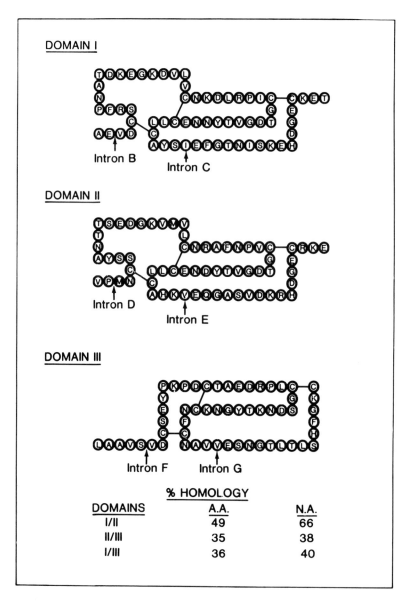

FIGURE 5.3. Alignment of the three domains of the secreted ovomucoid showing the similar positions of introns. *Top* portion shows the secondary structure of the three domains of the secreted ovomucoid polypeptide chain aligned for emphasis of the structural similarities. Disulfide bonds are indicated by *solid lines*. Positions of the six introns that interrupt the DNA sequence coding for the three domains are shown with *arrows*. Homology between the amino acid sequence of the domains and the mRNA sequence coding for these domains is shown at the *bottom*. (A.A. = amino acids; N.A. = nucleic acids.)

INTERVENING SEQUENCES IN MOLECULAR EVOLUTION

tions of the introns within the domains. Each domain is separated from the next by one intervening sequence. All domains also contain one internal intervening sequence at identical positions. As the lower part of Figure 5.3 shows 49 percent of the amino acids of domains I and II are identical. Sixty-six percent of the mRNA sequences coding for domains I and II are identical. The homology of domains I and II with domain III is less extensive but still significant. This homology suggests that the domains are related, a point that will be discussed in more detail later.

Although introns B, D, and F apparently separate the genomic domains coding for the functional peptide domains shown in Figure 5.3, what can be said about the other three introns—C, E, and G? Note that these three introns occur at identical positions within domains I, II, and III. These "internal" introns divide the subdomains that code for the trypsin-binding sites, the N-terminal halves of the peptide domains in Figure 5.3, from the C-terminal subdomains. It is possible that a primordial ovomucoid domain consisted only of the 5' subdomain and that the internal introns were created in the course of the addition of the 3' subdomain. The addition of another disulfide bond to the primordial ovomucoid polypeptide (a single domain) may have resulted in increased stability, increased rigidity of the active site, and/or resistance to digestion by the protease bound by the ovomucoid protein. If these subdomains were separated and the disulfide bridge broken, the functional integrity of the inhibitor would probably be destroyed. Thus, we made a strong argument that intervening sequences divide the ovomucoid gene into structural domains of the ovomucoid polypeptide that appear to correspond as well to functional polypeptide domains. We concluded that the ovomucoid intervening sequences specify functional domains of the protein and suggested that this is likely a general concept of eukaryotic gene evolution.

EVOLUTION OF THE OVOMUCOID GENE

The hypothesis that the present ovomucoid is the result of two intragenic doubling events is tenable in light of our investigation of the structural organization of the ovomucoid gene. Introns B, D, and F delineate three functional domains of the secreted ovomucoid polypeptide that each contain an internal intron at identical positions within the genomic coding sequence. A more detailed analysis of the DNA sequences surrounding the six splice points indicated in Figure 5.3 is even more revealing. The splice points (B, D, and F) that separate each domain occur between two codons of the genomic DNA sequence; in other words, these three introns are said to be in the same phase (phase O) with respect to the amino acid sequence of the protein. The internal splice points (C, E, and G) within each domain, however, occur between the second and third nucleotides of a codon of the genomic DNA sequence, and so these three introns all occur in phase II. Thus, the pattern of intron interruption of the codon sequence is identical both in position and phase. These observations are only plausible if the present ovomucoid gene is the result of two intragenic duplications of a primordial ovomucoid gene that contained an intron.

We would like to present now a likely scenario for the evolution of the ovomucoid gene. The primordial ovomucoid gene consisted only of the 5' subdomain of domain III. Intron G would have been created in the course of addition of the 3' subdomain, which would have added stability to the primordial protease inhibitor, as discussed earlier. This recombinational event would have resulted in the creation of the original, stable ovomucoid domain, represented as domain III in Figure 5.3. Two genomic duplications, with the retention of noncoding sequences within and at the 5' end of the primordial gene, would subsequently have led to the present ovomucoid gene. After the initial duplication event, in which domain III gave rise to domain II, 27 nucleotides (coding for nine amino acids; see Figure 5.3) must have been deleted in the 5' region of domain III. The duplication of domain II to form domain I must have occurred much more recently, as judged by the significant homology between the nucleotides at corresponding positions of the genomic sequences coding for domains I and II.

We can estimate an approximate time when this second duplication must have occurred by comparing the avian ovomucoids with mammalian protease inhibitors. The dog submandibular protease inhibitor has been sequenced and shown to consist of two polypeptide domains; domain II (the carboxy terminal domain) shows close homology to the ovomucoid carboxy terminal domain III (Kato et al., 1978). Because all avian species examined to date contain ovomucoids with three domains, the duplication to form the newest domain (I) must have occurred after the divergence of birds from mammals (about 300 million years ago). Therefore, the evolutionary ancestor to birds and mammals must have contained an ovomucoid with two domains, and we can conclude that the intervening sequences must have existed in the ovomucoid gene well before 300 million years ago.

Up to this point, our discussion has not included the signal peptide because it is not part of the secreted ovomucoid, and is, of course, significantly different from the three polypeptide domains. Intron B separates the genomic DNA coding for the bulk of the secreted ovomucoid from the genomic domain that codes for the signal peptide and the first two amino acids of the secreted ovomucoid polypeptide. However, intron A interrupts this coding sequence after the first 19 amino acid codons of the signal peptide. In our evolutionary scheme, the genomic ovomucoid domain containing these first two exons must have been added to the 5' end of the primordial ovomucoid at the time of the development of the capacity for secretion of the ovomucoid polypeptide.

Our theory for the evolution of the ovomucoid gene and its resultant protein structure is presented in Figure 5.4. The top line shows a single exon sequence, which coded for an ancestral protease inhibitor. This inhibitor was probably only moderately successful at its evolved task, the inhibition of proteases. However, with the addition of another appropriate exon to the 3' end of the original exon (step 1), the primordial ovomucoid gene was created, which coded for a stable inhibitor because of the addition of the strategically located third disulfide bond, as discussed earlier. This polypeptide corresponds to the present-day ovomucoid domain III. Step 2 represents the initial duplication of the

INTERVENING SEQUENCES IN MOLECULAR EVOLUTION 101

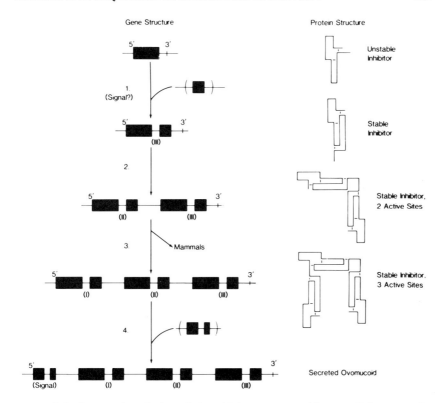

FIGURE 5.4. Proposed evolution of the chicken ovomucoid gene. Column on the *left* represents the ovomucoid gene structure at various times during its evolution; the protein structure for which it codes is shown in the column to the *right*. The 5' and 3' represent the transcription initiation and termination sites, respectively. *Solid bars* represent exon sequences. Between exons, the *thin lines* represent intervening sequences; outside of the exons, the *thin lines* represent untranslated or flanking sequences.

primordial ovomucoid gene, perhaps by unequal crossing over to form an ovomucoid gene coding for a stable inhibitor with two active sites, domains II and III. It was at this point in evolution, about 300 million years ago, when the avian and mammalian lineages diverged. Mammals have retained this two-domain protease inhibitor, whereas the avian gene evolved further. Step 3 represents the duplication of domain II, again most likely an unequal crossover event, to create a new domain, the present polypeptide domain I. The result was an ovomucoid gene that codes for the three-domain inhibitor that birds have today. Because all avian ovomucoids have this three-domain structure, the partial genomic duplication of step 3 must have occurred before the avian speciation, or 80–300 million years ago. Step 4 represents the addition of the exons coding for the signal peptide, which created an ovomucoid that could be secreted by the cells responsible for its synthesis (tubular gland cells of the oviduct). However, as noted in step 1, this addition could have occurred at any point in the evolution of the gene and, indeed, must have occurred before the secretion of the protease inhibitor became a necessity.

THE OVOINHIBITOR GENE

Recent results from the laboratory of one of the authors (B.W.O.) strengthens some of the hypotheses made in the previous sections concerning the role of introns in gene evolution and also supports our overall scheme for the evolution of the ovomucoid gene. In addition to ovomucoid, chickens contain at least three other members of the Kazal family of serine proteinase inhibitors (Laskowski and Kato, 1980). One of the others, ovoinhibitor (OI), is present in chicken egg white at about one tenth the level of ovomucoid, but it is a major proteinase inhibitor in chicken plasma (Liu et al., 1971). The chicken gene coding for ovoinhibitor was recently mapped to a region about 9–23 kb upstream from the 5' end of the ovomucoid gene, and its sequence organization has now been determined (Scott et al., 1987). The sequence of the chicken ovoinhibitor protein can be divided into seven structural domains, as shown in Figure 5.5. Each of these seven domains of the ovoinhibitor molecule is structually identical to the domains of chicken ovomucoid, since three disulfide bridges occur in identical positions. Each domain is also encoded by two exons. The sites at which the introns occur in the coding sequence of the domains are shown by the heavy black arrows in Figure 5.5. One intron separates the codon for the first amino acid of each domain from the preceding connecting peptide. In addition, one intron occurs at an identical position within the coding sequence of all the domains. These intradomain introns of the ovoinhibitor gene occur in the same phase (phase II) within the same amino acid codon as that in the ovomucoid domains. Furthermore, the last ovoinhibitor domain (VII), like the carboxy terminal domain III of ovomucoid, contains a small deletion within the region between the first two disulfide bridges (a "b"-type Kazal domain). The first six ovoinhibitor domains are "a"-type Kazal domains, just like domains I and II of the ovomucoid protein. The intermolecular amino acid sequence homology between the ovoinhibitor and ovomucoid domains varies from 28 to 54 percent. This similarity in domain organization and sequence between ovoinhibitor and ovomucoid implies that these proteins evolved from a common ancestor.

Since the intragenic "a"-type domain homologies are about the same as the intergenic "a"-type domain homologies, an exact duplication of a primordial ovomucoid gene to yield two inhibitor genes, one of which evolved into the present-day ovoinhibitor gene, could have occurred almost anywhere in our proposed evolutionary scheme of the chicken ovomucoid gene (see Fig. 5.4). The two "b"-type domains each contain deletions in an identical position, though they differ in extent. This implies that the region at the 5' end of these Kazal domains between the first two cysteines is not a critical region for function and that the deletions resulting in the present ovomucoid and ovoinhibitor "b"-do-

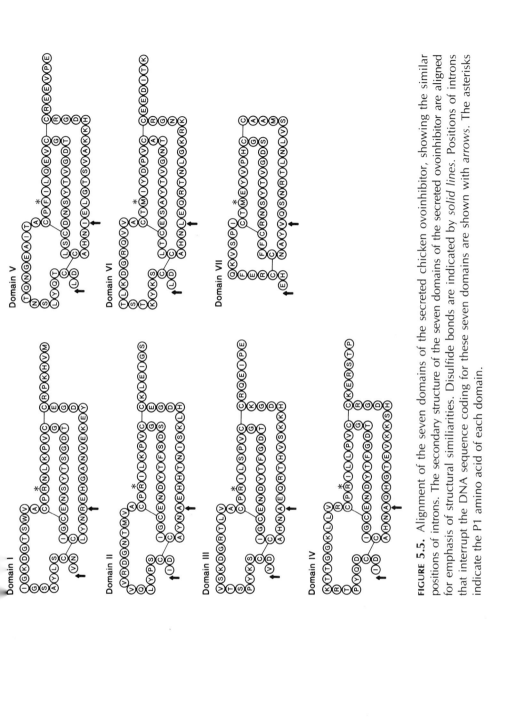

FIGURE 5.5. Alignment of the seven domains of the secreted chicken ovoinhibitor, showing the similar positions of introns. The secondary structure of the seven domains of the secreted ovoinhibitor are aligned for emphasis of structural similarities. Disulfide bonds are indicated by *solid lines*. Positions of introns that interrupt the DNA sequence coding for these seven domains are shown with *arrows*. The asterisks indicate the P1 amino acid of each domain.

main exon structures were necessarily independent events. These deletions must therefore have occurred after the original divergent gene duplication provided the precursors to the two proteinase inhibitor genes. Once this original gene duplication provided two independent copies of the primordial inhibitor gene, they evolved independently and the "a"-domain duplications, which led to the present inhibitor genes, were carried much farther in the ovoinhibitor gene than in the ovomucoid gene.

The repeated duplication of domain coding sequences would suggest that there was some selective pressure for domain duplication. It is the P1 amino acid of each domain, indicated with asterisks in Figure 5.5, that provides these independently functioning polypeptide domains with their proteinase specificity. Intramolecular domain duplication allows a single inhibitor molecule to develop the capacity to inhibit multiple proteinase molecules over the course of evolution. When coupled with codon changes at the P1 position of these multiple-domain exons, the inhibition of multiple types of proteinase molecules by one inhibitor molecule becomes possible. This provides a concrete example of how gene duplication provides an evolutionary means of creating diversity.

We previously suggested that the primordial Kazal inhibitor gene consisted only of the largest (5') exon of a domain, which encodes the reactive sites and five of the six conserved cysteines (Stein et al., 1980). The addition of the small (3') exon, which encodes the sixth cysteine, probably created a more efficient inhibitor as breakage of disulfide bonds tends to decrease inhibitor activity. This hypothesis is supported by the fact that the intradomain intron occurs in an identical location within the coding sequences of the seven ovoinhibitor domains as well as the three ovomucoid domains. Do these exons code for polypeptides that have a discrete structure or function?

The tertiary structures for many of these Kazal inhibitor domains are known (Bolognesi et al., 1982; Papamokos et al., 1982); they contain similar elements of secondary structure, a triple, standard antiparallel β-sheet, and an α-helix. The reactive sites, which have a similar geometry but are not part of the β-sheet or the α-helix, occur near the surface of the protein. In these inhibitors the α-helix is followed by a surface loop. An examination of the intradomain intron location reveals that the first exon of the ovomucoid or ovoinhibitor domain encodes the two long strands of the β-sheet and most of the α-helix. The polypeptide encoded by the second exon of the domain has little secondary structure other than a small third strand of the β-sheet. There would be, however, many hydrogen bonds formed between the polypeptides encoded by the two exons. Thus, the intradomain intron is similar to several of the chicken pyruvate kinase introns and a triphosphate isomerase intron, which fall between but not within gene sequences encoding stretches of α-helix (Lonberg and Gilbert, 1985; Marchionni and Gilbert, 1986). It seems possible, then, that the polypeptide encoded by the first exon can fold into a functional structure that is stabilized by interactions with the polypeptide encoded by the second exon. The structure of these Kazal inhibitors, therefore, is consistent with our hypothesis for the evolution of the ovomucoid (and ovoinhibitor) genes and with

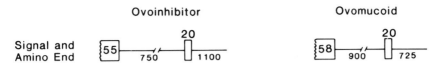

FIGURE 5.6. Structure of the peptide coding domains at the 5' end of the ovmucoid and ovoinhibitor genes. Exons are shown in *boxes*, with their length in nucleotides above or within the box. The intervening sequences are drawn as *lines*, with their length below.

Gilbert's and Blake's hypothesis that exons encode functional domains of proteins.

The sequence organization of the ovoinhibitor gene clarifies one additional variable in our scheme for the evolution of the ovomucoid gene. The structure of the peptide coding domains at the 5' end of the ovoinhibitor and ovomucoid genes are virtually identical, as shown in Figure 5.6. Both genes contain a 20-bp exon (exon II) that codes for the carboxy terminal portion of the signal peptide and the first two amino acids of the mature protein. The bulk of the signal peptide, which is one amino acid longer in ovomucoid than ovoinhibitor, is encoded by the 3' terminal portion of exon I in each gene. This conservation of genetic organization necessitates that the addition of the two exons that encode the signal peptide, shown as step 4 in the evolutionary scheme depicted in Figure 5.4, must have occurred before the duplication of the primordial inhibitor gene that led to the precursors of the present-day ovomucoid and ovoinhibitor genes. Since all of these Kazal inhibitors are secreted proteins, it is likely that the acquisition of the exons encoding the signal peptide would be an initial event in the formation of the primordial inhibitor gene, as denoted in step 1. Thus, our new evolutionary scheme for the ovomucoid gene would now divide step 1 in two: Step 1a would be the addition of the 3' subdomain exon to create the stable primordial ovomucoid domain (the old step 1), and step 1b would be the addition of the exon(s) encoding the signal peptide sequence (the old step 4). Furthermore, an additional duplication step is required subsequent to step 1a, in which the entire ancestral ovomucoid gene is duplicated, in order to supply an ancestral ovoinhibitor gene. This ancestral ovoinhibitor gene, containing two Kazal domains, then evolved independently through single domain duplications into its present seven domain structure.

In summary, the results of our sequence analyses of the chicken ovomucoid and ovoinhibitor genes, presented in the previous sections, provide support for the functional relevance of intervening sequences to gene structure. The adjacent position of these two genes on the same chromosome, as well as the extensive homology in the structural organization of the individual domains of the two genes, indicate that they both evolved from a common ancestral gene. The identity of intron location relevant to the polypeptide domain structure, as presented earlier, argues that both genes evolved via intragenic duplications of an ancestral gene coding for a single, functional domain. This analysis thus provides support for the hypotheses that exons encode structural–functional polypeptide domains and that introns existed at the time of (and were created by?) the assembly of primordial genes.

INTRONS AND INTRAGENIC DUPLICATION

There are now numerous examples of genes whose intron–exon organization suggests their evolution by successive intragenic duplication. Among these are the immunoglobin heavy-chain genes (Sakano et al., 1979), collagen (Yamada et al., 1980), TFIIIA (Tso et al., 1986), and the albumin/α-fetoprotein genes (Sargent et al., 1981; Eiferman et al., 1981). The albumin–α-fetoprotein gene pair, in particular, has several parallels with the ovomucoid–ovoinhibitor situation. These genes code for proteins that exhibit extensive sequence homology (32 percent). Based upon the recognition of a thrice-repeated pattern of disulfide bridges in the amino acid sequence of both human and bovine albumins, Brown first suggested that the ancestral gene itself had been formed by amplification and divergence of a simpler sequence (Brown, 1976). Several albumin and α-fetoprotein genes now have been sequenced; they contain numerous introns, but in all cases there is an excellent correlation of intron position within the three domains. The mouse albumin and α-fetoprotein genes are linked on chromosome 5, and sequence–structure homologies are consistent with their divergence from a common ancestral gene about 300–500 million years ago. Like the ovomucoid–ovoinhibitor family, studies on the binding sites of α-fetoproteins are consistent with the notion that the protein domains are functionally distinct and that the selective pressure to maintain amplified domains lay in the generation of functional diversity within the protein (Brown, 1976). In these few examples, investigators have proved that intragenic duplication, because of unequal crossingover within intervening sequences, is the principal evolutionary mechanism for the accretion of exons and introns. In view of the large number of proteins with internal periodicity, intragenic duplication must certainly have been an important evolutionary source of the complexity of eukaryotic genes.

EARLY ORIGINS OF INTERVENING SEQUENCES

The hypotheses briefly discussed in an earlier section, and indeed, our underlying assumption, require that introns have been present from the beginning of the organization of genetic information, as discrete structural–functional units, into genes. Analyses of the structure of the ovomucoid–ovoinhibitor and albumin–α-fetoprotein gene pairs, both of which evolved by intragenic domain duplications, led to the conclusion that introns were present in the primordial single-domain genes about 500 million years ago. This is inferential evidence, however, based solely on comparisons of present-day gene and protein structures. Is there any more direct evidence to support this hypothesis? Since it is unlikely that we will ever obtain fossilized remains of primordial organisms that would yield DNA for genetic analysis, a direct examination of primordial gene structure is precluded. However, in certain cases the examination of the same gene from different present-day organisms, presumably representing different stages in genetic evolution, has proved informative. One such opportunity occurred several years ago, when Gilbert and co-workers isolated and characterized two nonallelic, functional, rat preproinsulin genes (Lomedico et al., 1979). Although both genes showed extensive sequence homology, one (rI2) contained two introns, whereas the other (rI1) contained only one, in an identical position within the coding sequence to the 5' rI2 intron. The similarity of the two genes, coupled with phylogenetic evidence, suggested that they were the product of a gene duplication event that occurred during rodent evolution before the mouse–rat divergence (about 25–30 million years ago). The hypothesis of early origins of intervening sequences would argue that the rI1 gene had precisely lost the 3' intron, as opposed to the rI2 gene gaining the 3' intron by random insertion. This observation allowed the first test of the hypothesis: if a more ancestral preproinsulin gene were characterized (from a nonmammalian species), a structural comparison would show whether an old intron was lost or a new intron arose in the rI2 gene. Gilbert and co-workers then isolated the chicken preproinsulin gene (there is only one), which should represent the more ancestral preproinsulin gene in the form that existed before the rat preproinsulin gene divergence (Derler et al., 1980). The chicken gene indeed contained two introns, at positions identical to those of the rI2 gene introns, thus supporting the position that the ancestral preproinsulin gene contained two introns, one of which has been lost by one of the rat preproinsulin genes.

More recently, Gilbert and co-workers have added another, similar example to the argument that introns were present at the beginning of gene divergence. They examined the structural organization of several triphosphate isomerase (TIM) genes, which code for a highly conserved, ubiquitous glycolytic enzyme that most likely evolved before the divergence of eukaryotes and prokaryotes (Gilbert et al., 1986). Whereas *Escherichia coli* and *Saccharomyces cerevisiae* have uninterrupted TIM genes, the single chicken TIM gene contains six intervening sequences (Straus and Gilbert, 1985). Marchionni and Gilbert (1986) then isolated and sequenced the maize TIM gene (one of nine) and found eight introns. Of these eight, five were in positions identical to the chicken TIM gene; one was shifted by three codons; and two others, located near either terminus, were additional introns in maize. This "demonstrated" that the ancestral TIM gene was broken up at these positions before the divergence of plants and animals, which occurred about 1 billion years ago. Subsequently, the chicken gene has precisely lost two introns. McKnight et al. (1986) reported the sequence of the TIM gene from the filamentous fungus *Aspergillus nidulans*. They found five introns in this gene; two are found in the same position as two maize TIM gene introns, and two are unique to *A. nidulans*. One intron is found in an identical position in *Aspergillus,* maize, chicken, and man. The two introns uinque to *A. nidulans* are found in regions of the protein that were predicted to be interrupted by introns based on an analysis of a Go plot of the chicken TIM gene. These data now push back the evolutionary date to which introns can be inferred to exist within the ancestral TIM gene to before the evolutionary divergence of the filamentous fungi and higher eukaryotes, or about 1.2 billion years ago.

EXON SHUFFLING IN EVOLUTION

An additional implication of the hypothesis on the origin and function of intervening sequences presented in this chapter is that functional domains encoded by discrete exons would "shuffle" between different proteins, allowing the evolution of proteins as new combinations of preexisting functional units. Thus, we would expect occasionally to find short regions of unrelated genes that are homologous and composed of discrete exons. Recently, several potential examples of such "exon shuffling" have come to light. Three otherwise disparate enzymes—human phosphoglycerate kinase (PGK), maize alcohol dehydrogenase (ADH), and chicken glyceralde-3-phosphate dehydrogenase (GAPDH)—contain structurally similar 6-β-stranded nucleotide binding domains, which can be divided into two equivalent 3-β-stranded units (Michelson et al., 1985; Branden et al., 1984; Stone et al., 1985). These nucleotide binding domains, carboxy terminal in PGK and ADH, and amino terminal in GAPDH, are associated with structurally dissimilar catalytic domains. The genetic organization of the nucleotide binding domain does suggest a common origin for this region of these three genes. In each gene this domain is specified by five exons. Furthermore, in each gene the first mononucleotide subdomain is encoded by three exons, whereas the second mononucleotide subdomain is encoded by two exons. Thus, structural considerations lead to the conclusion that a primordial genetic unit, consisting of several exons and specifying a nucleotide binding domain, was combined with three structurally and functionally disparate genetic units that then evolved to form the present-day genes encoding the PGK, GAPDH, and ADH enzymes.

Exon shuffling has occurred during the creation of the serine protease gene family, which contains several proteins whose gene structure outside the protease coding domain is a pastiche of exons homologous to exons found in otherwise unrelated genes (Rogers, 1985). There are calcium-binding exons in the human clotting Factor IX (FIX) and thrombin genes, an exon homologous to an EGF exon in the FIX, human tissue plasminogen activator (TPA) and pig urokinase genes, and a single disulfide-linked "finger" exon in the TPA gene that is homologous to the tandemly repeated "fingers" of the fibronectin molecule.

Another well-documented case of exon shuffling occurs in the human LDL receptor. This protein contains several well-defined structural–functional domains, including a signal peptide, a cysteine-rich repeat region that contains the LDL binding domain, a large partially repetitive segment with sequence homology to the EGF precursor, a short region containing 18 clustered serine and threonine residues (many of which are carbohydrate-linked), a 22-amino-acid hydrophobic membrane-spanning domain, and a conserved C-terminal cytoplasmic domain (Sudhof et al., 1985a). The signal peptide and the three C-terminal domains are all encoded by one or two exons. More germane to this discussion are the two larger protein domains. The LDL binding domain is made up of seven repeats of 40 residues, each containing six cysteine residues spaced at similar intervals. The intron–exon pattern of this domain suggests that it arose by tandem duplications of a primordial exon. Each of the seven repeats is strongly homologous to a single 40-residue unit in complement factor C9, the only other gene in which this exon has been found to date. Following this LDL binding domain in the LDL receptor is a region of 400 amino acids that is 33 percent identical to a region of the EGF precursor. This region of both genes is encoded by eight exons that have apparently evolved from a common ancestral gene (Sudhof et al., 1985b). Also repeated three times within this region of both genes is an exon that encodes a 40-aa cysteine-rich sequence that is homologous to a repeat found in three proteins of the blood-clotting system: Factor IX, Factor X, and protein C. The occurrence of these shared sequences encoded by discrete exons in the LDL receptor gene lends credence to Gilbert's hypothesis that introns facilitate the evolution of new proteins from diverse combinations of preexisting functional units.

Our conceptual bias in writing this chapter is that introns began simply as spacer DNA regions surrounding primordial exons in which recombination events had occurred in the assembly of functional multiexon genetic units; and thus they are relics of gene creation via exon assembly. Introns themselves, however, have evolved into more complex structures, since many now contain various sequence recognition signals for splicing and transcription enhancers. We believe we have presented some cogent arguments for the functional relevance of introns to gene evolution, structure, and function. We have summarized evidence that introns indeed existed in the primordial precursors to some present genes, implying that introns are as old as the genes themselves. Introns seem to be disappearing from eukaryotic genes with time, and not increasing (except by intragenic duplications). The examples presented argue that exons began as functional elements of protein structure. Subsequently, exon shuffling and intragenic duplications have contributed greatly toward the creation of the modern "gene unit" through the assembly and sorting of elements of genetic information. Finally, we have shown how the structural analyses of genes, coupled with the structure–function analyses of proteins, provide a "window" through which we can observe the complex process of evolution.

ACKNOWLEDGMENTS

We would like to thank our colleagues who over the years have made major contributions to this research: Tony Means, Ming Tsai, Jim Catterall, Paula Kristo, Gene Lai, Clark Huckaby, and Michael Laskowski. We are especially grateful to Robert Schwartz for his encouragement and patience. Finally, we would like to acknowledge Ms. Lori Hogan for her assistance in the preparation of this manuscript.

References

Aloni, Y., Dhar, R., Laub, O., Morowitz, M., and Khoury, G. (1977). Proc. Natl. Acad. Sci. USA 74:3686–3690.
Berget, S. M., Moore, C., and Sharp, P. A. (1977). Proc. Natl. Acad. Sci. USA 74:3171–3175.
Blake, C. C. F. (1978). Nature 273:267.
Bolognesi, M., Gatti, G., Menegatti, E., Guarneri, M., Marquant, M., Papamokos, E., and Huber, R. (1982). J. Mol. Biol. 162:839–868.
Brack, C., and Tonegawa, S. (1977). Proc. Natl. Acad. Sci. USA 74:5652–5656.
Branden, C. I., Eklund, M., Cambillan, C., and Pryor, A. J. (1984). EMBO J. 3:1307–1310.
Breathnach, R., Mandel, J. L., and Chambon, P. (1977). Nature 270:314–319.
Brown, J. R. (1976). Fed. Proc. 35:2141–2144.
Catterall, J. R., Stein, J. P., Kristo, P., Means, A. R., and O'Malley, B. W. (1980). J. Cell. Biol. 87:480–487.
Derler, F., Efstratiadis, A., Lomedico, P., Gilbert, W., Kolodner, R., and Dodgson, J. (1980). Cell 20:555–566.
Doel, M. T., Houghton, M., Cook, E. A., and Carey, N. M. (1977). Nucleic Acids Res. 4:3701–3713.
Doolittle, W. F. (1978). Nature 272: 581.
Dugaiczyk, A., Woo, S. L. C., Lai, E. C., Mace, M. L., Jr., McReynolds, L., and O'Malley, B. W. (1978). Nature 274:328–333.
Eiferman, F. A., Young, P. R., Scott, R. W., and Tilghman, S. M. (1981). Nature 294:713–718.
Feeney, R. E. (1971). Proceedings of the First International Research Conference on Proteinase Inhibitors, Fritz, H., and Tschesche, H. eds. Walter De Gruyter, Berlin, pp. 162–168.
Feeney, R. E., and Allison, R. G. Evolutionary Biochemistry of Proteins (1969). Wiley Interscience, New York.
Gilbert, W. (1978). Nature 271:501.
Gilbert, W., Marchionni, M., and McKnight, G. (1986). Cell 46:151–154.
Glover, D. M., and Hogness, D. S. (1977). Cell 10:167–176.
Jeffreys, A. J., and Flavell, R. A. (1977). Cell 12:1097–1108.
Kato, I., Kohr, W. J., and Laskowski, M. J., Jr. (1978). Proc. FEBS Meeting 47:197–206.
Lai, E. C., Woo, S. L. C., Dugaiczyk, A., and O'Malley, B. W. (1979a). Cell 16:201–211.
Lai, E. C., Stein, J. P., Catterall, J. F., Woo, S. L. C., Mace, M. L., Means, A. R., and O'Malley, B. W. (1979b). Cell 18:829–842.
Laskowski, M. J., Jr., and Kato, I. (1980). Ann. Rev. Biochem. 49:593–626.
Lineweaver, H., and Murray, C. W. (1947). J. Biol. Chem. 171:565–572.
Liu, W.-H., Means, G. E., and Feeney, R. E. (1971). Biochim. Biophys. Acta 229:176–185.

Lomedico, P., Rosenthal, N., Efstratiadis, A., Gilbert, W., Kolodner, R., and Tizard, R. (1979). Cell 18:545–588.
Lonberg, N., and Gilbert, W. (1985). Cell 40:81–90.
Marchionni, M., and Gilbert, W. (1986). Cell 46:133–141.
McKnight, G. L., O'Hara, P. J., and Parker, M. L. (1986). Cell 46:143–147.
Michelson, A. M., Blake, C. C. F., Evans, S. T., and Orkin, S. M. (1985). Proc. Natl. Acad. Sci. USA 82:6965–6969.
Papamokos, E., Weber, E., Bode, W., Huber, R., Empie, M. W., Kato, I., and Laskowski, M. J., Jr. (1982). J. Mol. Biol. 158:515–537.
Rogers, J. (1985). Nature 315:458–459.
Sakano, H., Rogers, J. M., Muppi, K., Brack, C., Traunecker, A., Maki, R., Wall, R., and Tonegawa, S. (1979). Nature 277:627–633.
Sargent, T. D., Jagodzinski, L. L., Yang, M., and Bonner, J. (1981). Molec. Cell. Biol. 1:871–883.
Scott, M. J., Huckaby, C. S., Kato, I., Kohr, W. J., Laskowski, M. J., Jr., Tsai, M.-J., and O'Malley, B. W. (1987) J. Biol. Chem. 262:5899–5907.
Stein, J. P., Catterall, J. F., Kristo, P., Means, A. R., and O'Malley, B. W. (1980). Cell 21:681–687.
Stone, E. M., Rothblum, K. N., and Schwartz, R. J. (1985). Nature 313:498–500.
Straus, D., and Gilbert, W. (1985). Molec. Cell. Biol. 5:3497–3506.
Sudhof, T. C., Goldstein, J. L., Brown, M. S., and Russell, D. W. (1985a). Science 228:815–822.
Sudhof, T. C., Russell, D. W., Goldstein, J. L., Brown, M. S., Sanchez-Pescador, R. and Bell, G. I. (1985b). Science 228:893–895.
Tso, J. Y., Van Den Berg, D. J., and Korn, L. J. (1986). Nucleic Acids Res. 14:2187–2200.
Weinstock, R., Sweet, R., Weiss, M., Cedar, M., and Axel, R. (1978). Proc. Natl Acad. Sci. USA 75:1299–1303.
Yamada, Y., Avvedimento, V. W., Mudryg, M., Ohkubo, H., Vogeli, G., Ivani, M. Pastan, I., and de Crombrugghe, B. (1980). Cell 22:887–892.

6

Different Types of Introns and Splicing Mechanisms

PHILIP S. PERLMAN,
CRAIG L. PEEBLES,
AND CHARLES DANIELS

Intervening sequences in genes were originally described in 1977 when they were detected in the major late transcription unit of adenovirus (Berget et al., 1977; Chow et al., 1977). Those reports demonstrated that the functional mRNA contained stretches of sequence information that had been separate in the viral DNA. In the mRNA, those sequences were joined together as a continuous polynucleotide. The sequence blocks in the mRNA were arranged in the same order, 5' to 3', as they had been in the DNA, with the intervening sequences omitted. Soon it became clear that this unexpected gene structure was not limited to adenovirus, but was a general characteristic of eukaryotic genes (reviewed by Abelson, 1979; Padgett et al., 1986).

As the DNA sequence data base has grown during the past decade, the number of examples of genes with intervening sequences has grown in parallel. The hypothesis that intron removal occurs at the level of RNA processing (rather than by some alternative process such as DNA rearrangement or transcriptional skipping) has now become firmly established (see Crick, 1979; reviewed by Padgett et al., 1986). Introns have been found in genes encoding most major classes of RNA products. The vast majority of introns have been found in nuclear genes coding for proteins, but some have been described for other nuclear genes, including some rRNA and tRNA genes. Organellar and viral genomes also have many examples of introns, and a few have been discovered in certain bacteriophages and archaebacteria.

It is now clear that introns are not all alike, even in the most general terms.

Four distinct groups of introns are currently recognized as having distinct structural features and processing requirements; two of those groups have already been divided into subgroups. Additional intron groups are likely to be discovered as molecular analysis is applied to the genes of other types of organisms. Further subdivisions of the major groups will probably be defined as our understanding of the details of splicing mechanisms within each group becomes more detailed.

This volume is chiefly concerned with the largest group of introns, those found in nuclear pre-mRNAs, and most of the chapters deal with functional and evolutionary consequences of the presence of introns in nuclear genes of higher eukaryotes. The four groups of introns and the manner in which each is spliced are discussed in this chapter. The evolution of nuclear pre-mRNA introns is thoroughly considered elsewhere in this volume; therefore, that topic will not be discussed here. We will present more evolutionary considerations of the three other groups of introns.

THERE ARE FOUR DISTINCTLY DIFFERENT TYPES OF INTRONS

Nuclear Pre-mRNA Introns

Introns have been found in at least some nuclear genes encoding proteins of all eukaryotes so far examined. A large number of such introns have been identified; a recent compilation of splice-junction sequences included nearly two thousand examples (Shapiro and Senapathy, 1987). Nuclear pre-mRNA introns vary in the frequency of their occurrence within genes of different species. Great variation of intron size exists and some organisms have introns with distinctive average base compositions. A typical vertebrate protein-encoding gene usually has several introns of 200–500 nt, and almost all such genes contain at least one intron. By contrast, only a minority of genes in the yeast *Saccharomyces cerevisiae* contain any introns, and those that do usually have only one intron. There is also significant variation of intron size among various organisms; for example, typical introns of the nematode *Caenorhabditis elegans* are substantially shorter than those found in vertebrates (Blumenthal and Thomas, 1988). In *Drosophila melanogaster* most introns are shorter than 200 nt (e.g., Falkenthal et al., 1985), but several extremely long ones, over 50 kb, have been found (e.g., Karch et al., 1985; Chen et al., 1987). Plant introns are relatively A+T-rich (e.g., Goodall and Filipowicz, 1989) but are otherwise similar in size to those of vertebrates. The longest intron reported to date is in the Duchenne muscular dystrophy gene and is over 200 kb in length (Monaco et al., 1986).

The conserved sequence elements of nuclear pre-mRNA introns are short and primarily define the boundaries; little of the internal sequence is required beyond maintenance of a certain minimum length. In 1978, Breathnach et al. noted that the boundaries of the introns of the ovalbumin gene have short conserved sequences (Fig. 6.1A). While they examined only six introns, the con-

A.   5' -- UCAG/GUA   -----   UNCAG/G - 3'
              5' SJ   3' SJ

B.   5' -- AG/GUAAGU   --   CURA*Y   -   (C/U)$_{10+}$ NCAG/G - 3'
              5' SJ   BRANCHSITE   3' SJ

C.   5' -- /GUAUGU   --   UACUAA*C   -   (C/U)AG/   --   3'
              5' SJ   BRANCHSITE   3' SJ

D.

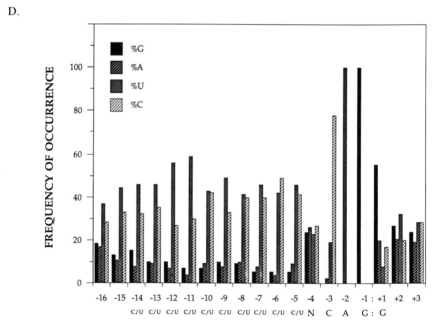

POSITION RELATIVE TO 3' SPLICE JUNCTION

FIGURE 6.1. Intron boundaries of nuclear pre-mRNAs are conserved sequences. (A) Breathnach et al. (1978) examined the boundaries of the introns from the chicken ovalbumin gene and identified this early form of the consensus for the splice sites. Note that their consensus was developed from a sample of just six introns and without knowledge of the precise positions of the processing events from direct

E.

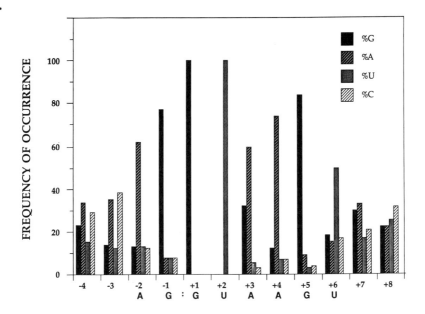

POSITION RELATIVE TO 5' SPLICE JUNCTION

RNA analysis. (B) Conserved splicing sites for vertebrate genes are displayed as the consensus sequences according to the compilation of Padgett et al. (1986) and confirmed by Shapiro and Senapathy (1987). (C) Conserved splicing sites determined for *S. cerevisiae* genes are an almost invariant special case of the general consensus and are shown essentially as described by Teem et al. (1984) and compiled in Padgett et al. (1986). (D) Graphical presentation of the sequence preferences within the consensus at the 3' splice junction for vertebrate introns reveals that certain positions are much more variable than others. These data are from the compilation by Padgett et al. (1986) and include only those examples that end at an -AG- dinucleotide. Note that the pyrimidine preference becomes more evident as the 3' splice junction is approached and that U is slightly more frequent than C at most of those positions. Also note that the positions -4, $+2$, and $+3$ show virtually no preference for any of the four nucleotides. Finally note that the preference for C at -3 is much stronger than the preference for G at $+1$. (E) Graphical presentation of the sequence preferences within the consensus at the 5' splice junction for vertebrate introns identifies certain positions as more strongly conserved than others. These data are from the compilation of Padgett et al. (1986), and that compilation only included examples of introns that started with the -GU- dinucleotide. The strong preferences for G at -1, A at $+4$, and G at $+5$ are consistent with analyses that show severe effects of mutations at these positions on splicing. The preferences for A at $+3$ and -2 are weaker, whereas the preference for U at $+6$ is not strong. Either A or C is slightly preferred at both positions -3 and -4, whereas there is no discernible preference for a particular nucleotide at positions beyond $+6$.

sensus they identified was basically correct. The presently accepted consensus sequence for splice junctions has been refined and extended somewhat, owing to the increased sample size and the inclusion of examples from a wide range of eukaryotic organisms (Breathnach and Chambon, 1981; Mount, 1982; Brown, 1986; Green, 1986; Padgett et al., 1986; Shapiro and Senapathy, 1987) (see Fig. 6.1B).

The averaging process inherent in developing the consensus, however, obscures recognition of any distinctions between types of organisms or subgroups of introns. There is good evidence for such distinctions in at least a few cases. For example, the 5' end of a typical intron is defined by the consensus sequence GURAGU; yet, in yeast it is specifically GUAUGU (Teem et al., 1984) (see Fig. 6.1C). Other examples of organism-specific 5' boundary sequences have yet to be identified; however, this one case reveals that the complexity of the splicing process and strategies for attainment of specificity vary among eukaryotes. Introns of yeast genes also use a specific sequence, UACUAAC, at the branch point (Langford and Gallwitz, 1983; Pikielny et al., 1983). By contrast, animal and (probably) plant introns use the less discriminating consensus, CURAY; and only rarely have the yeast sequence (Keller and Noon, 1984). Yeast introns are distinct in another way from the vertebrate introns; they lack the pyrimidine-rich sequence between the branch-site and the conserved AG at the 3' junction. The 3' boundaries of plant introns also differ somewhat from those of animals (compare Brown, 1986 with Mount, 1982; cf., Shapiro and Senapathy, 1987).

Early studies revealed that introns from one mammal are readily spliced by another. On the other hand, baker's yeast generally does not splice introns from animals (Beggs et al., 1980; Langford et al., 1983; Watts et al., 1983). This probably reflects the special sequences used for the branch-site and 5' splice junction region in yeast. The fission yeast, *Schizosaccharomyces pombe,* can splice at least some introns from vertebrate genes, suggesting that this microbe has a splicing machinery more like that of animals (Kaufer et al., 1985). Introns from vertebrates and plants have similar conserved sequences, yet vertebrate introns are not efficiently spliced in plant cells (Barta et al., 1986; van Santen and Spritz, 1987; Wiebauer et al., 1988). Some plant introns are accurately spliced in HeLa cells or extracts (e.g., Brown et al., 1986; Hartmuth and Barta, 1986), while others are not (Wiebauer et al., 1988). Where heterologous systems fail to splice correctly, the problem is usually with 3' site selection (e.g., Wiebauer et al., 1988).

There are only a few short sequences present in essentially all nuclear pre-mRNA introns; those sequences are recognizably similar among all eukaryotes and have been summarized as the consensus sequences for the splice junctions and the branch site. As described in more detail in the section titled Nuclear Pre-mRNA Introns, under Splicing Mechanisms, those sequences play important roles in the splicing mechanism by interacting with trans-acting splicing factors. Since those sequences are located at or near the intron boundaries, they are presumably the main determinants of site specificity. The fact that the splice-site sequences have been retained in similar forms and roles throughout the

DIFFERENT TYPES OF INTRONS AND SPLICING MECHANISMS 117

eukaryotic kingdom implies that these nuclear introns are all evolutionarily related. However, there are organism-specific sequence preferences for certain sites and several cases where one organism cannot correctly splice an intron from another organism. It is probably important for protein evolution that the conserved sites for splicing these introns are restricted to the boundaries. That is, recombination within an intron would not usually destroy it as a substrate for splicing. Likewise, the splice sites are autonomous for the most part, meaning that one 5' splice junction can be utilized with various 3' splice junctions (and vice versa).

Introns of tRNA Genes

The early finding that genes for tyrosine (Goodman et al., 1977) and phenylalanine (Valenzuela et al., 1978) tRNAs from the yeast *Saccharomyces cerevisiae* are interrupted by short sequences not found in the mature, functional tRNAs established that splicing is not limited to nuclear protein-encoding genes of higher organisms and their viruses. One of the genes sequenced in that early investigation was known to be functional in vivo, as it was a nonsense-suppressor allele (Goodman et al., 1979). These results showed that production of mature tRNAs required removal of the extra sequences, and posttranscriptional RNA processing seemed to be the likely mechanism. It was also immediately apparent that more than one mechanism has to exist to accomplish RNA splicing, since the introns in pre-tRNA genes are much smaller than and lack the boundary sequences associated with nuclear pre-mRNA introns. It was natural to suppose that tRNA splicing was a different process, since tRNAs have a conserved and highly base-paired structure, while mRNAs do not exhibit a conserved secondary structure as a group.

Introns have been identified in nuclear tRNA genes from a broad range of eukaryotic organisms (Guthrie and Abelson, 1982), but only a subset of tRNA genes contain introns. This contrasts with the frequent occurrence of introns in protein-encoding genes from vertebrate organisms. Examination of the nuclear tRNA gene sequences that have introns reveals that each contains only a single small intron of 8–60 nucleotides. A catalog representing over thirty distinct examples is shown in Fig. 6.2. Such interrupted nuclear tRNA genes have been identified in fungi, plants, and vertebrates. The genes for tRNAs are usually repeated and form gene families for particular isoacceptors. If an intron is present, it is found in all members of that particular family.

Transfer-RNA genes of chloroplast DNA are sometimes interrupted by group I or group II introns but not by introns similar to those of nuclear tRNA genes (e.g., Koch et al., 1981; Steinmetz et al., 1982). Introns have been identified in some archaebacterial tRNA genes, as well (Kaine, 1987; Kaine et al., 1983; Daniels et al., 1985; Wich et al., 1987). These introns will be discussed separately, as they are spliced by different mechanisms.

Introns of nuclear tRNA genes always interrupt the mature sequence at the same site (Fig. 6.3). These introns are one nucleotide away from the 3' side of the anticodon, next to the conserved purine that is usually hypermodified in the

#	tRNA Identity AA-AC/Source/	tRNA 5'-Half Molecule Sequence	Intron Sequence	tRNA 3'-Half Molecule Sequence
1	ILE-UAU/Sce/	GCTCGTGTAGCTCAGTGGTTAGAGCTTCGTGCTTATA	GCAACATTCGGTTCCGAAGTTCTGCCAAAGACCCTTTCAAACAGCCCTTTAAAACA	ACCCCACGGTCGTGGGTTCAATCCCACTCACGACGA
2	ILE-UAU/Sce/	GCTCGTGTAGCTCAGTGGTTAGAGCTTCGTGCTTATA	GCAACATTCGGTTCCGAAGTTCTGCCAAAGACCCTTTCAAACAGCCCTTTAAAACA	ACCCCACGGTCGTGGGTTCAAACCCCACTCACGACA
3	LEU-CAA/Dme/	GTCAGGATGGCCGAGCGGTCTAAGGCGCCAGACTCAAG	ATTGAAAATCTTACTTTCTGAACGTTTGTGTTTGTTAATGAGCG	TTCTGGGTCTCTGTGAGCGTGGGTTCGATTCCCACTCATCTGAC
4	LEU-CAA/Dme/	GTCAGGATGGCCGAGCGGTCTAAGGCGCCAGACTCAAG	ATTTAAAATCTTACTTTCTGAACGAAAGTGTATGAGCG	TTCTGGGTCTCTGTGAGCGTGGGTTCGATTCCCACTCATCTGGA
5	TRP-CCA/Sce/	GAAGCGGTGGCTCAATGGTAGAGCTTTCGACTCCAA	TTAAATCTTGAAATTCACGGAATAAGATTGCA	ATTGAAGGTTGCAGGTTCGAGTCCTGCTCGGTTTCA
6	LEU-CAA/Sce/	GGTTGTTTGGCCGAGCGGTCTAAGGCGCCTGATTCAAG	AAAAAATCTTGACCGCCAGTTAACTGTGGGAATA	CTCAGGTATGTAAGATCGAAGATTGCAATCTTCAGCAAACCA
7	LEU-CAA/Sce/	GGTTGTTTGGCCGAGCGGTCTAAGGCGCCTGATTCAAG	AAATATCTTGACCGAGTTAACTGTGGGAATA	CTCAGGTATGTAAGATCGAAGATTGCAATCTTCAGCAAACA
8	PRO-UGG/Sce/	GGGCGTGTGGTCTAGTGGTATGATTCTCGCTTTGGG	CGACTTCCTGATTAAACAGGAAGACAAAGCA	TGCAGAGGCCCCGGGTTCAATTCCCGGCTCGCCC
9	PRO-UGG/Sce/	GGGCGTGTGGTCTAGTGGTATGATTCTCGCTTTGGG	CGACTTCCTGCTAAACAGGAAGACAAAGCA	TGCAGAGGCCCCGGGTTCAATTCCCGGCTCGCCCC
10	PRO-UGG/Sce/	GGGCGTGTGGTCTAGTGGTATGATTCTCGCTTTGGG	CGACTTCCTGCCTAAACAGGAAGACAAAGCA	TGCAGAGGCCCCGGGTTCAATTCCCGGCTCGCCC
11	PRO-UGG/Sce/	GGGCGTGTGGTCTAGTGGTATGATTCTCGCTTTGGG	CTGTGAAATAAACAGGAAGACAAAGCCA	TGCGAGAGGCCCCGGGTTCAATTCCCGGCTCGCCC
12	LEU-AAG/Ncr/	GCCAAGATGGCCGAGCGGTCTAAGGCGCCACGTTAAG	TTAACCCTTAATATTCCTTCCAAGGTT	CCGTGGTCCGAAAGGCCGTGGGTTCGAACCCACCATCTTGGAC
13	LYS-UUU/Sce/	TCCTTGTTAGCTCAGTTGGTAGAGCGTTCGGCTTTTA	AGCGCATTTCGTAAGCAAGGAT	ACGAAGTGCAGGGTTCGAGTCCCCTTCAACAAGAC
14	TYR-GUA/Hsa/	CCTTCGATAGCTCAGCTGGTAGAGCGGAGGACTGTAG	ACTGCGGAAACGTTGTGGAC	ATCTTAGGTCGCTGGTTCGAATCCAGCTCGAAGGA
15	TYR-GUA/Hsa/	CCTTCGATAGCTCAGCTGGTAGAGCGGAGGACTGTAG	ATTGTACAGACATTTGCGAC	ATCTTAGGTCGCTGGTTCGAATCCAGCATTTGCA
16	TYR-GUA/Hsa/	CCTTCGATAGCTCAGCTGGTAGAGCGGAGGACTGTAG	CTACTTCCTCAGCAGGAGAC	ATCTTAGGTCGCTGGTTCGAATCCAGCATTTAAGGA
17	LEU-UAG/Sce/	GGAGTTTGGCCGAGTGGTTTAAGGCGTCAGATTTAGG	TGGATTTAACCTCTAAAT	CTCTATATTCGAATGCAAGGTTCGAATCCTTGTCAACTCCA
18	LEU-UAG/Sce/	GGAGTTTGGCCGAGTGGTTTAAGGCGTCAGATTTAGG	TGGGTTTAACCTCTAAAT	CTCTATATTCGAATGCAAGGTTCGAATCCTTGTCAACTCCA
19	PHE-GAA/Sce/	GCGATTTAGCTCAGTTGGGAGAGCGCCAGACTGAAG	AAAAACTTCGGTCAAGTT	ATCTGAGGTCTGTGTTCGATCACACAGAATTCGCA
20	SER-CGA/Sce/	GGCACTATGGCCGAGTGGTTAAGGCGAGAGACTGCAA	TGGAATAAAAAAGTTCGCT	ATCTTCGGGTCTCCCCGTCGAATCCCGTCTAGTGCCACCA
21	SER-GCU/Sce/	GTCCCAGTGGCCGAGTGGTTAAGGCGATGCCCTGCTA	TTTCCTCAGAAAAGCAATT	AGCAATGGGTTTTACCCGGCGAGGGTTCGAATCCCTCTGGGACA
22	PHE-GAA/Sce/	GCGGATTTAGCTCAGTTGGGAGAGCGCCAGACTGAAG	AAATACTTCGGTCAAGTT	ACTCGAGGTCTGTGTTCGATCACACAGAATTCGCA
23	SER-UGA/Pan/	GTCAGCATGGCCGAGTGGTCTAATGCGTTAGACTTGAA	TATCCATTACATTCAGT	ATCTAATTCCTCGGCGACGTTAGTTCGAACTAACGTTGCCTGAC
24	LEU-UAA/Spo/	GGGCTATGCCCGAGTGGTCTAAGGGGCAGATTTAAG	AGGCCTCGGCCTTGTA	CCTTCGTGTCTGTAAACGGAGATTAGTTGCAATCTAATCAGGGA
25	PHE-GAA/Ncr/	GCGGTTTAGCTCAGTTGGGAGAGCGTCAGACTGAAG	TCCACTTCACTCATAA	ATCTGAAGTGCGCTCGGGCCGCCAGTTCGAATCTGGCCAACGCA
26	SER-CGA/Spo/	GTCACTATGTCCGAGTGGTTAAGGAGTTAGACTGAA	TTCCTACATTCGTGC	ATCTAATGCCCTCGGGCGCCAGTTCGAATCTGGCTAGTGACGAC
27	SER-UGA/Spo/	GTCACTATGTCCGAGTGGTTAAGGAGTTAGACTGAA	TCCTGTATTCTAGTC	ATCTAATGCGCTCGGGCCGCCAGTTCGAATCTGGCTAGTGACAC
28	TYR-GUA/Sce/	CTCTCGGTAGCCAAGTTGGTTTAAGGCGCAAGACTGTAA	TTTATCACTACGAA	ACACTTCGTCACGGTTCGAATCCGTAGCCCGGGTT
29	TYR-GUA/Sce/	CTCTCGGTAGCCAAGTTGGTTTAAGGCGCAAGACTGTAA	TTTACCACTACGAA	ACTTCGCATGGTCGCGTTCGAATCCCGCGCCCGGGTT
30	TRP-CCA/Dd1/	GACTCCTTAGCATAGTGGTTTATTGTAATTGTCTCCAA	AACGTTAGAAGTT	ACACTCGTCCAACGGTTCGAATCCCTTAAGGTGTCA
31	TYR-GUA/Nru/	CCGACCTTAGCTCAGTTGGTAGAGCGGAGGACTGTAG	TGGTACTGCTCAG	ATCTTAGGTGCCAGCGTTCGAATCACTGGAGGTCGGA
32	TYR-GUA/Xle/	CCTTCGATAGCTCAGCTGGTAGAGCGGAGGACTGTAG	GTGTGATCGAGCA	ATCTTAGGTCGCTGGTTCGAATCCAGCAGAGAAGGA
33	TYR-GUA/Xle/	CCTTCGATAGCTCAGCTGGTAGAGCGGAGGACTGTAG	AGGAATATAGCA	ATCTTAGGTCGCTGGTTCGAATCCAGCTCGAAGGA
34	MET-CAU/Gma/	GGGGTGGTGGCCAGTTGGCTAGCCGGTAGGTCTCATA	GCTTCTGAGTT	ATCACAGTTTGCAGTTCGATTCTGCAATCACCCCA

mature tRNA. Structure predictions based on free-energy calculations and structure-probing studies support the straightforward hypothesis that the secondary structure of pre-tRNA is essentially the same as the mature tRNA, but with an extra domain in a distinctive location (Swerdlow and Guthrie, 1984; Lee and Knapp, 1985). As a rule, naturally occurring introns of pre-tRNA genes contain a sequence that is complementary to the anticodon of the tRNA. Pre-tRNA introns are usually represented as basepaired to the anticodon and forming a duplex extension stacked on the anticodon stem. This duplex section is flanked by single-stranded segments that include the splice junctions. This is in contrast to all other classes of introns in which the intron location is generally not limited to a recognizable structural domain of the product RNA.

The processing mechanism of pre-tRNA gene introns apparently takes advantage of this special constraint on their location. Mutations that alter the basic structure of the mature domain or the spatial relationship between the mature and intron domains disturb the processing system in terms of rate, specificity (site selection), or both. This shows that the processing enzymes identify the correct sites by recognition of elements of the tRNA structure. Most point mutations and small deletions within the intron have little effect on splicing (e.g., Strobel and Abelson, 1986; Greer et al., 1987).

←

FIGURE 6.2. A compilation of DNA sequences for nuclear pre-tRNA genes containing introns. Each wildtype gene sequence is listed separately if there are any differences, even when the mature tRNA product is the same. Mutant forms and informational suppressor alleles are not included. This listing is displayed in order of decreasing intron length. Only eukaryotic nuclear genes are listed; additional classes of introns in tRNA genes are found in archaebacteria and chloroplast genomes. The anticodon (AC) is underlined and is also listed as part of the tRNA identifier at the left. The amino acid specificity (AA) is abbreviated with standard three-letter codes. The organism from which a gene has been isolated is abbreviated with a three-letter code derived from the initial letter of the genus and the initial two letters of the species designation as follows: *Drosophila melanogaster* (Dme), *Dictyostelium discoideum* (Ddi), *Glycine max* (Gma), *Homo sapiens* (Hsa), *Neurospora crassa* (Ncr), *Nicotiana rustica* (Nru), *Podospora anserina* (Pan), *Saccharomyces cerevisiae* (Sce), *Schizzosacharomyces pombe* (Spo), and *Xenopus laevis* (Xle). References for these sequences are as follows: (1) and (2) Ogden et al. (1984); (3) and (4) Robinson and Davidson (1981); (5) Kang et al. (1980); (6) Venegas et al. (1979); (7) Kang et al. (1980) and Andreadis et al. (1982); (8) and (9) Ogden et al. (1984); (10) Winey et al. (1986); (11) Lee and Knapp (1985); (12) Huiet et al. (1984); (13) del Rey et al. (1982); (14) and (15) MacPherson and Roy (1986); (16) van Tol et al. (1987); (17) and (18) Ogden et al. (1984); (19) Valenzuela et al. (1978); (20) corrected from Etcheverry et al. (1979), Olson et al. (1981), and Ogden et al. (1984); (21) Stucka and Feldman (1988); (22) Valenzuela et al. (1978); (23) Debuchy and Brigoo (1985); (24) Sumner-Smith et al. (1984); (25) Selker and Yanofsky (1980); (26) Mao et al. (1980); (27) Hottinger et al. (1982); (28) and (29) Goodman et al. (1977); (30) Peffley and Sogin (1981); (31) Stange and Beier (1986); (32) Muller and Clarkson (1980); (33) Gouilloud and Clarkson (1986); (34) Waldron et al. (1985); and (35) Gamulin et al. (1983).

120   INTERVENING SEQUENCES IN EVOLUTION AND DEVELOPMENT

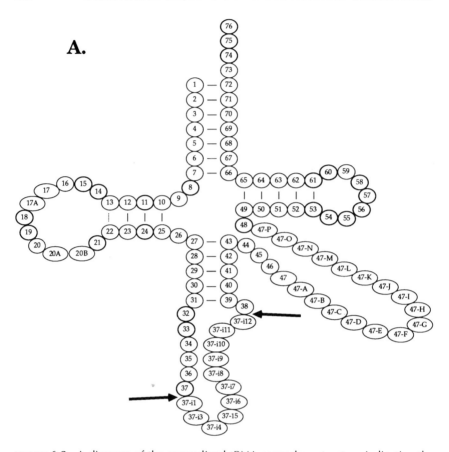

FIGURE 6.3. A diagram of the generalized tRNA secondary structure indicating the location of tRNA introns. (A) The generalized tRNA structure is extended to accommodate introns. Our own convention has been used to designate the intron nucleotides as 37-i*n*, where *n* is the nucleotide number from the start of the intron. In the example shown, an intron length of 12 nucleotides has been arbitrarily selected. The positions of the splice junctions for nuclear pre-tRNA introns are identified by the large arrows between nucleotides 37 and 37–i1 and between 37–i12 and 38. (B) Relative to the mature tRNA, additional intron locations are indicated for pre-tRNA introns found in archaebacteria and chloroplasts. These drawings are modeled after Figure 1 of Sprinzl et al. (1987), and that compilation has been used to enumerate the intron sites. Abbreviations are *CHLORO* for "chloroplast" and *AR-CHAE* for "archaebacteria." The anticodon positions are shaded for emphasis.

Examination of the sequences within tRNA introns does not reveal stretches of conserved boundary or internal sequences. There is a tendency for these introns to be rich in adenosine residues, but that does not appear to be functionally significant since an artificial intron consisting of a stretch of poly(U) can still be accurately and efficiently removed (Reyes and Abelson, 1988). It should be noted that Reyes and Abelson detected an apparent requirement for

DIFFERENT TYPES OF INTRONS AND SPLICING MECHANISMS

B.

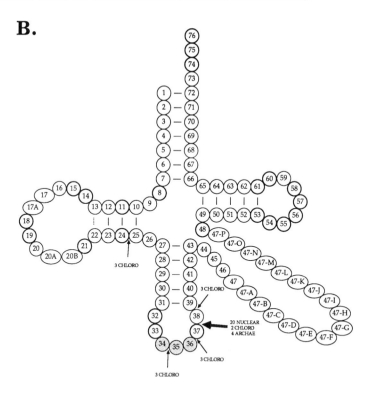

a purine preceding the 5' splice junction. All natural tRNAs have a purine at that location. The most highly conserved nucleotide within tRNA introns noted to date is the third residue from the 3' end. It is a G nearly 80% of the time, but mutation of that site has little effect on splicing (Szekely et al., 1988).

In cases where the intron is large, a section of it can form a self-contained stem-loop structure known as the intron extra arm. Mutations constructed within such introns are not usually deleterious to splicing, unless the splice sites become embedded in double-stranded segments or the overall structure of the tRNA domain is disturbed. Insertions can be made in the intron variable arm, for example, to increase the size of the intron far above the currently observed natural maximum of sixty nucleotides, with little effect on splicing (Johnson et al., 1981; Raymond and Johnson, 1983). However, insertions that are very stably base-paired internally, or mutations that result in base-pairing to the splice junctions, can block splicing entirely (Willis et al., 1984; Greer et al., 1987; Szekely et al., 1988). The consensus is that an intron can be spliced as long as it is sufficiently compact to "fit" into the active site of the cleavage enzyme with the splice sites in single stranded loops.

The introns of yeast pre-tRNAs have been studied the most intensively. Some nine families of yeast tRNA genes contain introns (Ogden et al., 1984). Since most tRNAs are encoded by multiple genes, there are many more than nine genes with introns of this type. For a particular tRNA isoacceptor, all genes have at most one intron. Within each gene family, the introns are all very similar (if not identical) in sequence. In the case of pre-tRNA phenylalanine, for example, there are genes with both 18- and 19-nucleotide-long introns; but the two intron sequences are almost identical. This may reflect some process of sequence rectification that acts to maintain homogeneity within the gene family. Between tRNA gene families, however, introns vary substantially in sequence (Ogden et al., 1984).

This argues that there is little in the way of specific recognition of the intron sequence and provides further support for the idea that the processing activities depend primarily on structural features of the mature domain to identify the splice junctions. Apparently there is a single splicing endonuclease activity in yeast and other organisms that can recognize and process all of the distinct families of pre-tRNAs. This also argues that most of the important recognition elements are contained within the tRNA domain (Otsuka el al., 1981; Peebles et al., 1983). These features are also conserved throughout the eukaryotic lineage, since yeast pre-tRNA genes can be accurately transcribed and processed in vitro in heterologous systems derived from amphibian or mammalian cells (Ogden et al., 1979; Standring et al., 1981; DeRobertis et al., 1981).

Since most tRNA genes lack introns, it is not immediately obvious why any have been retained. They are probably not preserved for a purpose such as exon shuffling, since the base-paired structure of a tRNA makes swapping exons pointless. One possible explanation is suggested by experiments that involve removing the intron from suppressor alleles of tRNA genes. A gene so modified produces a functional version of the suppressor tRNA but it is significantly less effective than the suppressor made from the intron-containing gene. Analogous results have been reported for both tyrosine- and leucine-specific genes of yeast. In both cases, the reduced suppressor function was traced to a lack of a specific base modification within or near the anticodon loop (Johnson and Abelson, 1983; Strobel and Abelson, 1986). Only the intron-containing precursor serves as the substrate for the modifying enzyme. Recently, it has been shown that the genes for tyrosine-specific tRNAs from humans also contain introns (McPherson and Roy, 1986; van Tol et al., 1987) and that removal of the intron from the gene reduces suppressor function by preventing the same nucleoside modification as in the yeast system (Tol and Beier, 1988). Similar studies of a tyrosine tRNA from *Drosophila* resulted in the same conclusions (Suter et al., 1986; Choffat et al., 1988). Thus, one reason for retention of tRNA introns is a requirement for the pre-tRNA as a modification enzyme substrate. This device was probably fixed in evolution before the divergence of major groups of eukaryotes. It is not known whether this hypothesis is sufficient to explain the persistence of all known nuclear tRNA gene introns.

Figure 6.3 also indicates the locations of tRNA gene introns in archaebacteria and in chloroplasts. Some of these are located at the same site as the nuclear tRNA introns, but others have distinctive locations. Most chloroplast tRNA introns are group II introns. Several intron-containing cpDNA tRNA genes have been sequenced from two or more plants. In each case the intron was at the same location and was highly conserved in sequence. One of these introns is very large, about 2500 bp, and contains a free-standing open reading frame capable of encoding a protein. A few chloroplast tRNA genes have group I introns less that 1000 bp long. The archaebacterial tRNA gene introns are quite a bit shorter than those, but some are significantly longer than the longest known nuclear tRNA intron.

Group I Introns

As early as 1977, researchers studying the structure and function of genes in yeast mitochondrial DNA (mtDNA) used genetic data to reach the conclusion that at least one such gene has interrupted coding sequences (Slonimski et al., 1978; Haid et al., 1979; Hanson et al., 1979; Alexander et al., 1980). This was entirely unanticipated because it was already dogmatic that mitochondrial and chloroplast DNAs were derived from prokaryotic endosymbionts of primitive nucleated organisms. Since prokaryotes were not then known to have introns, confirmation of introns in yeast mtDNA was accepted only after the publication of direct physical and sequence data (Bonitz et al., 1980; Dujon, 1980; Nobrega and Tzagoloff, 1980).

By late 1980, the primary sequences of eleven mitochondrial introns of yeast had been published. It was clear to those in the field that boundary sequences typical of nuclear pre-mRNA introns were absent from mitochondrial introns. Between 1979 and 1983 a number of other "atypical" intron sequences were found in such diverse places as nuclear rDNA of *Tetrahymena thermophila* and *Physarum polycephalum,* mitochondrial and chloroplast DNA of various plants, and mitochondrial DNA of other fungi (reviewed by Cech et al., 1983; Michel and Dujon, 1983; Waring et al., 1983).

In an important paper published in 1982, Michel et al. stated clearly that there are two distinctly different kinds of "atypical" introns in organelle genomes. The two kinds are now called group I and group II. In the following year, it was recognized that group I introns are present in nuclear DNA of at least two lower eukaryotes (Cech et al., 1983; Michel and Dujon, 1983; Waring et al., 1983). More recently, group I introns have also been found in several structural genes of the bacteriophages T2, T4, T6, and SPO1 (Chu et al., 1984; Gott et al., 1986; Pedersen-Lane and Belfort, 1987; Sjoberg et al., 1987; and reviewed in Shub et al., 1987). An intron apparently related to the group I introns has even been described in the large rRNA gene of an archaebacterium (Kjems and Garrett, 1985). Group I introns are commonly found in large rRNA genes (both nuclear examples, a number in mtDNA of fungi, cpDNA of *Chlamydomonas* species, and the archaebacteria); several examples of chloroplast tRNA genes interrupted by group I introns are known; but, most group I introns are found in protein-encoding genes.

Group I introns have been investigated intensively by genetic and biochemical approaches. Three groups have studied splicing-defective mutants of intron 4 of the cytochrome b gene of yeast mtDNA (or bI4) and noted that *cis*-acting mutations are clustered in short sequences that are conserved among group I introns (Anziano et al., 1982; de la Salle et al., 1982; Weiss-Brummer et al.,

A.

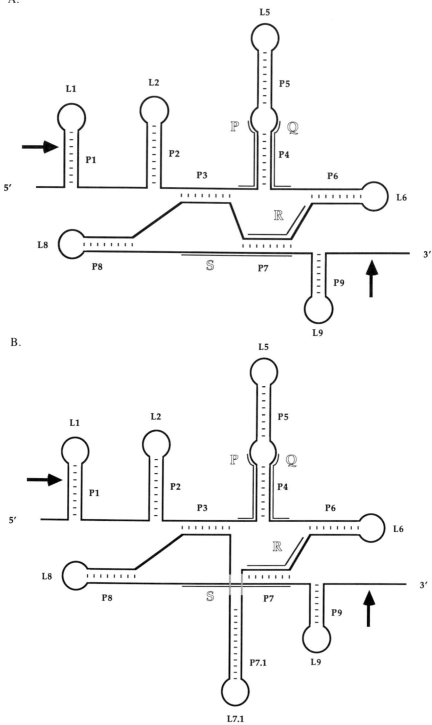

B.

DIFFERENT TYPES OF INTRONS AND SPLICING MECHANISMS

1982). Davies et al. (1982) and Michel et al. (1982) independently developed the hypothesis that group I introns have conserved RNA secondary structures. The conserved sequences previously shown to be functionally important for splicing proved to be involved in defining specific, conserved, structural interactions.

Figure 6.4 is a diagram of the conserved core structure of group I introns (adapted from Burke et al., 1987). It is made up of nine helices (or pairings), nine loops, and flanking 3' sequences. There are several subgroups. In one, P2 is absent; and, in another there is an extra helix, P7.1. Put another way, in the majority of group I introns, the 5' sequence of P7 is adjacent to the 3' sequence of P3. Where helix P7.1 is present, those parts of P7 and P3 are separated by an insertion that makes up P7.1 plus L7.1. Consensus sequences for the elements of P4 and P7 are shown in Fig. 6.5. The rest of the pairings involve sequences that are not conserved. It should be added that the last base of the 5' exon is always a U in group I introns and the last base of the intron is always a G. In fact, provided that the exon/intron boundaries are known, group I introns can be distinguished from group II or nuclear pre-mRNA introns simply by examining the boundaries. More lengthy internal conserved sequences and the resulting secondary structure models are largely confirmatory in making assignments. Recently, Kim and Cech (1987) have extended such modeling from two to three dimensions and described a rationale for testing new predictions of their model.

Davies et al. (1982) first proposed that an intron sequence that can pair with the 3' end of the upstream exon and the 5' end of the downstream exon is always present. This sequence is usually near the 5' boundary and has been termed the "internal guide sequence," or IGS, since it appears capable of aligning the two exon boundaries for splicing. Subsequent research has shown clearly that the pairing between the IGS and the 5' exon is important for splicing—and that interaction forms P1 in the conserved secondary structure (Fig. 6.4) (e.g., Been and Cech, 1986; Perea and Jacq, 1985). However, the pairing with the 3' exon is, as yet, unsupported by experimental evidence. Since P1 involves exon sequences that are not conserved except for the U residue at the exon/intron boundary, it follows that the IGS for each group I intron has a different sequence.

←

FIGURE 6.4. Secondary structure of the group I introns. The core secondary structure model of group I introns is represented in this schematic according to the conventions proposed by Burke et al. (1987). (A) Structural diagram for the majority subtype, Group IB introns, which includes the *Tetrahymena thermophila* LSU-rRNA intron. (B) Structural diagram for the minority subtype, Group IA introns, which include the bacteriophage T4 *td* intron. Pairings (Pn) are numbered from 5' to 3'. Loops (Ln) are generally designated by the number of the pairing that they enclose. (The exception is the loop between P4 and P5.) The locations of conserved sequences designated P, Q, R, and S, following the nomenclature of Davies et al. (1982), are indicated by the outline characters. The locations of the splice sites are shown by the bold arrows.

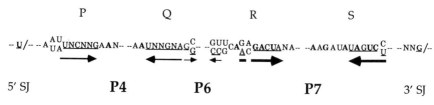

FIGURE 6.5. The consensus sequence elements for group I introns. Two key pairings that define the core secondary structure of Group I introns involve the interaction of the conserved primary sequence elements. P pairs with part of Q to form P4 and a portion of R pairs with S to form P7. Like pairs of arrows underscore the sequences involved in the pairings. The most strongly conserved nucleotides within this consensus are underlined. The 5' and 3' splice junctions (5'SJ and 3'SJ) are indicated with a slash mark (/).

Interestingly, the IGS pairing with the 5' exon almost always extends through the first few bases of the intron (as shown in Fig. 6.4). We have noticed that the first few bases of each group I intron are often repeated at the beginning of the 3' exon. Thus, the full pairing of the P1 structure competes, in many cases, with the "guide" structure proposed by Davies et al. (1982). While the pairing between the IGS and the 5' exon is supported by strong experimental tests, the possible role of the first few intron bases pairing with the IGS has not yet been tested. The apparent duplication of a short sequence at the flanks of these introns may be relics of the initial intron acquisition events.

Excellent in vivo and in vitro data, obtained using several introns, support the main aspects of the proposed core structures. For example, some in vivo mutations of yeast mitochondrial introns define functionally important sequences; more recently, some in vivo data are available for group I introns of bacteriophage T4 (e.g., Hall et al., 1987). The greatest wealth of mutant data is available for derivatives of the intron of the nuclear large rRNA gene of *Tetrahymena* (Price and Cech, 1985; Price et al., 1985, 1987; Waring et al., 1985; Burke et al., 1986; Williamson et al., 1987). As described in the next section, studies of that self-splicing intron defined the mechanism of splicing used by group I introns. Such studies have confirmed what was already known for other introns and extended the testing of the functional importance of structures other than P3 and P7.

Several examples of second-site suppression of primary defects in the core structure have been reported. In vivo experiments of yeast bI4 by Schweyen's group confirmed the pairings that form P3 and P7 (Weiss-Brummer et al., 1983; Holl et al., 1985). Extensive studies of that sort using in vitro mutants of the *Tetrahymena* intron confirm the functional importance of P1 (noted above), P3 (Williamson et al., 1987), and P7 (Burke et al., 1986). There remains little doubt that group I introns are highly structured RNAs and that their ability to splice is defined by that structure.

An important early observation concerning group I introns is that many of them contain sequences that encode a protein (intron open reading frames, or

ORFs). A clear example of an intron-encoded protein is the ribosomal protein S-5 encoded by the large rRNA gene intron of *Neurospora* mtDNA (Burke and RajBhandary, 1982). The analogous intron of yeast mtDNA encodes a different protein that is a sequence-specific DNA endonuclease involved in a special form of recombination (Colleaux et al., 1986). At least three yeast group I mitochondrial introns (bI2, bI3 and bI4) encode maturase proteins needed for splicing (see below). Kotylak et al. (1985) reported genetic evidence suggesting that the bI4-encoded protein also plays a role in recombination. And finally, the reading frame of yeast *coxI* intron 4 encodes a potent DNA endonuclease and a latent maturase (Delahodde et al., 1989; Wenzlau et al., 1989). Very recently, studies in a number of systems have shown that a DNA endonuclease is a common intron-encoded function (reviewed by Lambowitz, 1989). It must be stressed that the majority of group I intron ORFs have not yet been shown to encode a protein having demonstrable function. It seems likely that there are additional intron-encoded functions still to be discovered. This will be easiest when genetic methods can be applied, as these have proved to be most useful in defining the known intron-encoded functions.

Intron ORFs are usually in register with the reading frame of the preceding exon (e.g., Bonitz et al., 1980; Nobrega and Tzagoloff, 1980; Lazowska et al., 1980). When such ORFs are known to be expressed, they appear to be translated as fusion proteins encoded by both exon and intron sequences. In some cases, this translation product is processed proteolytically; both the yeast bI4- and aI4-encoded proteins are processed to separate the intron- and exon-encoded regions (Anziano et al., 1982; Hanson et al., 1982; Banroques et al., 1986). In contrast, the yeast bI2 maturase appears to function without such processing (Lazowska et al., 1980).

There are some examples of "free-standing" intron ORFs. The ORFs in the mitochondrial rRNA gene introns of *N. crassa* and *S. cerevisiae* are expressed (Burke and RajBhandary, 1982; Macreadie et al., 1985) and the excised introns are relatively stable (Tabak et al., 1984; Green et al., 1981), unlike some other group I introns in these organisms (Conrad, 1987; Collins and Lambowitz, 1985). It is tempting to conclude that the excised intron is the mRNA for the intron-encoded protein. Unfortunately, there is no definitive evidence of this in *Neurospora;* and, in yeast, one cannot distinguish between the excised intron or some RNA obtained from percursor RNA by an alternative processing (not splicing) pathway (Zhu et al., 1987 and 1989).

A different situation exists for group I intron ORFs from bacteriophage T4. These group I introns have free-standing open reading frames that can be expressed. However, it appears that they are translated from an mRNA that results from a promoter within the intron, rather than from the precursor RNA or an RNA excised from it (Gott et al., 1989). It is now likely that two of the three T4 intron ORFs encode endonucleases involved in intron mobilization like those in yeast mitochondria (Quirk et al., 1989).

When the core structures of ORF-containing group I introns are modeled, most of each ORF is found sequestered in a single substructure in nearly every case; but, the location of the coding region varies widely among the introns of the group. For example, among the introns with free-standing ORFs, the ORF is located in L6a for the T4 nrdB and td gene introns, in L9.1 for the T4 sunY and *N. crassa* strain 74A ND1 gene introns, and in L8 of the *S. cerevisiae* 21S rRNA gene intron. For introns with reading frames that are continuous with that of the upstream exon, similar variability of location of most of the reading frame is found. For example, L1 contains all of the reading frame of aI3 in yeast mitochondria while L8 contains most of the reading frame in yeast aI4 and bI4 (see Fig. 6.4 for a diagram of the core structures locating these loop regions). When an intron ORF is in frame with the preceding exon, and located anywhere other than in L1, some *cis*-acting sequences are part of the reading frame. However, it is clear for aI4 and bI4 that the functionally important parts of their ORFs are present in L8 and that the ORF portion preceding L8 is probably unimportant for function (e.g., Delahodde et al., 1989).

It has been suggested that group I intron ORFs may have been acquired relatively recently, perhaps by transposition or gene conversion events. If so, the site of insertion should influence the manner of expression of the ORF. As seen in the introns of phage T4, the ORFs are expressed from a separate promoter that forms an mRNA species beginning quite near the start codon of the free-standing ORF. For rRNA gene introns it appears that the pre-rRNA is not the mRNA for the intron-encoded protein but that some processed form of the pre-rRNA is translated. If the simplest way to swap reading frames is to have each ORF supply its own promoter, then the rRNA examples may have lost the promoter and then evolved a strategy to compensate.

The common situation where the intron ORF is fused to the exon reading frame may have resulted from an alternative adaptation to the loss of the promoter: the sequence between the exon and the beginning of the ORF had to be converted into triplets so that translation of the pre-mRNA results in translation of the intron ORF. Interestingly, there are at least two cases where an intron protein is made as a precursor and is processed proteolytically to release an active form of the intron-encoded protein. The system for processing such hybrid proteins is another example of a compensation for the loss of an intron promoter.

Group II Introns

Michel et al. (1982) were the first to point out that fungal mtDNAs contain a second, less common, type of intron. They noted that the distinguishing features of group I introns are absent from that second type, termed *group II*. Group II introns have a different set of short conserved boundary sequences (indicated in Fig. 6.6) and a longer sequence located near the 3' end of the intron (see Keller and Michel, 1985). The 5' conserved sequence has the consensus GUGCG; however, unlike group I introns, there is not even a single conserved nucleotide in the 5' exon. The 3' conserved sequence is the dinucleotide AU or AC; there are no exceptions to this. As noted above, if the location of one or both boundaries is known, inspection of a few nucleotides

```
5' - /GUGCG ---- RAGCYGUAUNNNRNGAAANUNNNACGUACRGUUY - A - A(U/C)/ - 3'
        ──▶       ─────▶                       ◀────    ◀──
```

5' SJ   Domain 5   3' SJ

FIGURE 6.6. Conserved primary sequences typical of group II introns. The consensus boundary sequences are shown along with a conserved internal sequence, domain 5. The 5' splice junction (5'SJ) and 3' splice junction (3'SJ) are indicated by the slash mark (/). The site of branch formation is an A residue (A*) located seven or eight nucleotides from the 3'SJ within a characteristic secondary structure, domain 6. The consensus sequence for domain 5 includes two helical segments, indicated by the two sets of similar arrows underscoring the paired positions. Domain 5 is the most highly conserved primary sequence of group II introns and is located within about 80 nucleotides of the 3'SJ.

there is sufficient to distinguish group II from other types of introns with a very high degree of confidence.

Group II introns have a conserved core secondary structure, but one that has nothing that is obvious in common with the core of group I introns. The group II core is represented as a central wheel with six spokes (helices) radiating from it (Fig. 6.7, and Jacquier and Michel, 1987). Today these helices are said to delineate substructures or "domains" of the introns. The highly conserved internal sequence is domain 5.

In the Splicing Mechanisms section, the splicing mechanism of group II introns is described and similarities to that of nuclear pre-mRNA introns are discussed. The adenosine residue involved in RNA branch formation of group II introns is located very close to the 3' end of each intron within domain 6 (van der Veen et al., 1986; Schmelzer and Schweyen, 1986). Two subgroups of these introns have been defined and the location of the branch A residue is one of the key distinguishing features. The branch A is the seventh nucleotide from the end in group IIA and the eighth in group IIB (F. Michel, personal communication).

While many group I introns have ORFs, only a minority of group II introns can encode a protein. Most of those are listed by Michel and Lang, 1985. Each intron ORF of a protein-encoding gene is in-frame with the preceding exon. There is an intron of a cpDNA tRNA gene (present in at least two plants) in which the ORF is free-standing (Sugita et al., 1985; Neuhaus and Link, 1987). Group II intron ORFs all encode large proteins, from 450 to over 700 amino acids long, and in all cases the majority of the coding capacity is sequestered within intron domain 4. It is interesting that nearly all group II intron ORFs contain a region of strong homology with retroviral reverse transcriptases (Michel and Lang, 1985). In the two cases (in yeast) where a function has been assigned to products of group II intron ORFs, they are maturases needed for splicing (Carignani et al., 1983; Mecklenburg, 1986).

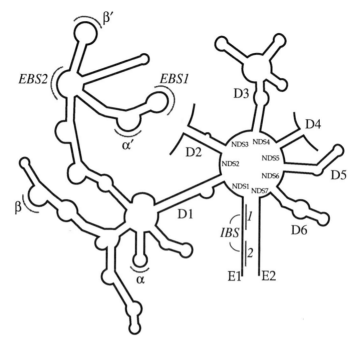

Figure 6.7. The core secondary structure of group II introns. The schematic diagram depicts the core secondary structure model for group II introns proposed by Michel and Dujon (1983). Sections of the structure are identified as domains (Dn) bounded by inverted repeat sequences. Nucleotides between domains are designated as nondomain segments (NDSn), conserved in length and somewhat conserved in sequence. Pairing interactions have been identified involving intron binding sites (IBS1 and IBS2) within the 5' exon (E1) and exon binding sites (EBS1 and EBS2) within D1 of the intron (Jacquier and Michel, 1987). Two additional interactions are indicated (α:α' and β:β'). This drawing is modeled after Figure 4 of Jacquier and Michel (1987).

SPLICING MECHANISMS

Nuclear pre-mRNA Introns

Most knowledge of the mechanism of nuclear pre-mRNA intron splicing comes from in vitro experiments using extracts of yeast cells or HeLa cell nuclei (reviewed by Green, 1986, and Padgett et al., 1986). Some requirements for specific intron sequences were detected earlier using site-directed mutagenesis and transient expression in cultured cells or transformation of yeast. It is clear that these introns require a number of trans-acting factors for their splicing, including proteins and small nuclear RNPs (snRNPs)—specific complexes of proteins and snRNAs. These accessory factors interact to form large complexes now known as *spliceosomes* and much current research has been devoted to defining the components and their order of assembly.

The pathway of splicing for nuclear pre-mRNAs requires the prior assembly of the spliceosome. Within this structure, two ordered reactions rearrange the pre-mRNA substrate, leading to the release of the spliced exons and the excised intron (Fig. 6.8). The first step is believed to be a transesterification (phosphodiester rearrangement) involving the transfer of the 5' phosphate from the first nucleotide of the intron (a guanosine) to the 2' hydroxyl of the ribose of the adenosine residue at the branch site near the 3' end of the intron. The products of the first step are the free 5' exon with a 3' hydroxyl and the IVS-3' exon in the form of a branched circular molecule termed a *lariat*.

The second step of splicing is believed to be a transesterification involving the phosphodiester preceding the first nucleotide of the downstream exon. This 5' phosphate is transferred to the 3' hydroxyl of the first exon to form the spliced exon product (Fig. 6.8). The phosphate at the splice junction is derived from the 5' phosphate of the first nucleotide of the 3' exon. The excised intron ends with a 3' hydroxyl, and the phosphate that forms the 2'–5' bond at the branch is derived from the 5' phosphate of the first intron nucleotide. After the second step, the splicing products separate and spliceosome is disassembled. The intron is turned over, and the spliced exons are either subject to more splicing or are exported to the cytoplasm as mature mRNA. There is no net change in the number of phosphodiester bonds in the RNA substrate as it is converted to these products. However, many bonds will be broken in the turnover of the intron.

The idea that these steps are transesterifications is based on the analogy to group I and group II intron self-splicing reactions that will be discussed later. Before self-splicing was discovered, it was thought that nuclear pre-mRNA splicing probably involved distinct endonuclease and ligase steps (as later found to be the case for tRNA intron splicing; see the following discussion). In favor of transesterification is the lack of evidence for the existence of a linear form of the IVS-3' exon intermediate associated with the spliceosome. Any cleavage and ligation pathway would include linear intermediates with free termini. Likewise, there is no evidence for a free form of the 3' exon associated with the spliceosome. If there were phosphomonoester intermediates at either step of splicing, there would have to be a way to activate and join them, as well. Transesterification avoids that difficulty. The 3' hydroxyl serves as a nucleophile to attack the phosphodiester directly. The free energy of the bond is preserved and no unobserved intermediates are predicted.

The spliceosome assembly process serves to select and align the splice junctions and branch site for the reaction. The assembly process occurs on each intron of a multi-intron transcript, demanding that all necessary sites of suitable polarity be identified prior to cleavage of any bond. The components of the spliceosome stay associated with the intron product primarily and are then disassembled to be reused in a new cycle.

Good evidence is available that this pathway operates in vivo. Branched RNA was identified in poly(A+) RNA from HeLa cell nuclei by isolating the nuclease-resistant trinucleotide following digestion of the RNA (Wallace and Edmonds, 1983). Others have identified lariat intermediates and intron products

A.

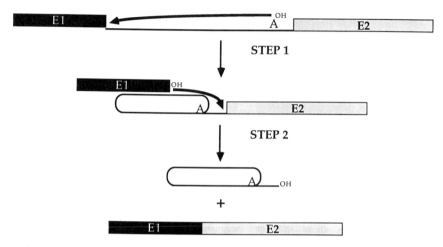

FIGURE 6.8. Schematic diagram of the reaction sequences for splicing, drawn to emphasize the similarity of three major pathways. (A) Splicing of pre-mRNA introns is depicted as a sequence of two transesterification reactions. In the first reaction (STEP1), the 2'OH of the branchpoint adenosine (A^{OH}) acts as a nucleophile to attack the phosphodiester at the 5'SJ, releasing E1 (dark shading) and forming the intermediate lariat containing the intron (solid line) and E2 (light shading). In the second reaction (STEP2), the 3'OH of E1 acts as a nucleophile to attack the 3'SJ, completing intron lariat excision and exon joining. This model for the chemical mechanism of nuclear pre-mRNA splicing also applies to group II self-splicing. (B) Splicing of Group I introns is depicted as two transesterification reactions. In the first reaction (STEP1), a guanosine mononucleotide cosubstrate (G_{OH}) acts as a nucleophile to attack the 5'SJ, releasing E1 (dark shading) and adding to the 5' end of the intron (solid line). In the second reaction (STEP2) the 3'OH of E1 acts as a nucleophile to attack the 3'SJ, joining the exons and releasing the intron as a linear RNA with the attached G. This description of splicing is most clearly articulated by Cech and Bass (1986).

from cellular RNA of mammalian tissue (Zeitlin and Efstratiadis, 1984) and from yeast cells (Domdey et al., 1984). In mammals, a particular branch site is used; but when it is altered, splicing can proceed using alternative, cryptic, branch sites (e.g., Ruskin et al., 1985). In yeast, where the natural branch site is always located within the sequence UACUAAC, some mutations of the site block splicing completely in vivo (e.g., Langford et al., 1984; Jacquier et al., 1985).

Most work has been done either with whole cell extracts of yeast cells (Lin et al., 1985) or with nuclear extracts of HeLa cells (Hernandez and Keller, 1983; Grabowski et al., 1984; Krainer et al., 1984). In these systems exogenous RNA is processed efficiently and accurately. The reaction requires modest concentrations of monovalent cations (usually potassium), divalent cations (usually

B.

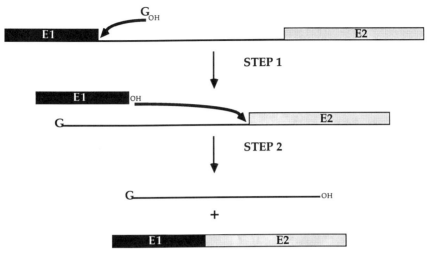

magnesium), and ATP. Often the system is supplemented with an ATP-generating system. ATP analogs with nonhydrolyzable bonds cannot support splicing, and other nucleoside triphosphates (except for dATP) are likewise ineffective. The extracts are generally prepared to be quite concentrated, and this makes good sense, since the process requires the assembly of a complex from many components. The reaction in extracts of vertebrate cells proceeds only after a significant time lag of 20 minutes or more. Again this is reasonable if a complicated assembly process is necessary for the first step in splicing to occur. The lag is not apparent, however, in the yeast version of this system. The earliest detectable products are the free 5' exon and the lariat IVS-3' exon RNA. Later, the spliced exon product accumulates and the free intron lariat appears.

The ATP is needed at several steps, and it probably participates in several different processes. Since there is no net change in the number of phosphodiester bonds in the RNA products relative to the precursor if the rearrangements are truly transesterifications, ATP is probably not required to activate RNA intermediates for ligation. Instead, the role of ATP could be as an energy-producing cofactor for RNA helicases that may act to remove secondary structure that otherwise would interfere with spliceosome assembly. Another role could be as a component of an assembly intermediate, where its hydrolysis would be coupled to conformational changes in the components. This would make such steps proceed in the forward direction exclusively. Still another possibility is that ATP hydrolysis could be used for mechanical work in assembling or disassembling the spliceosome or for conformational changes during the steps of splicing itself. While a simple tracking model for seeking splice sites may not be operative, mechanical coupling may still be necessary for the splicing process or for ancillary activities like RNA transport that might normally be coupled to splicing in pre-mRNA metabolism. ATP utilization may

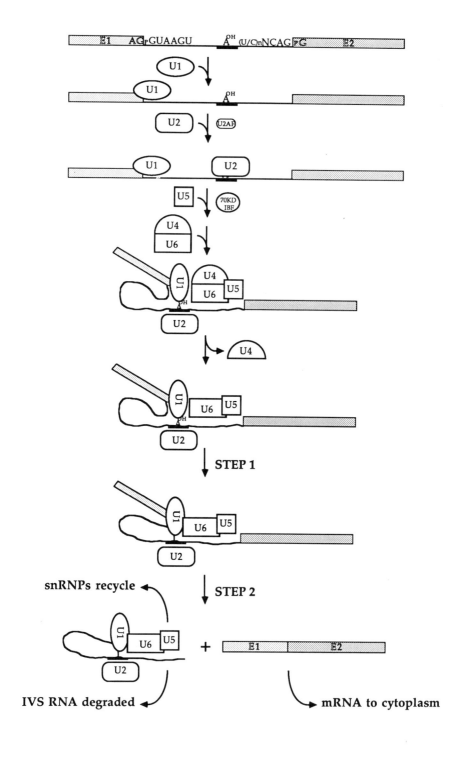

also be necessary to provide a driving energy for processes that would be unlikely to proceed spontaneously by themselves, such as removal of hnRNP proteins from sites needed for binding other factors.

The mechanism of pre-mRNA splicing is intimately connected with the spliceosome assembly process. The original evidence for the existence of spliceosomes came from gradient centrifugation of complexes formed between model pre-mRNAs and components present in extracts needed for splicing (Frendeway and Keller, 1985; Grabowski et al., 1985; Brody and Abelson, 1985). It was found that the pre-mRNA associates with rapidly sedimenting (40–60S) complexes. Assembly requires ATP, magnesium ion, and a functional pre-mRNA. The spliceosome fraction from the gradient contains mostly substrates and intermediates of splicing. Some final products are also present, but not in stoichiometric amounts. This suggests that the products are released and complexes dissociate after splicing.

The gradient centrifugation technology has been supplanted by native gel electrophoresis methods (Konarska and Sharp, 1986; Pikielny, et al., 1986) and high-resolution gel filtration and related chromatographic methods (Grabowski and Sharp, 1986; Perkins et al., 1986; Reed and Maniatis, 1988). Those studies revealed that the assembly process yields an initial spliceosome complex that contains intact precursor RNA. The first covalent change in the substrate follows (Fig. 6.9). Then the spliceosome is modified to a new form with a distinct mobility on the gel systems; it contains mostly splicing intermediates.

The complexes contain many components in addition to the precursor RNA and reaction intermediates. Direct evidence exists for the presence of snRNAs as well as several protein factors, including hnRNP proteins, as essential components. Both biochemical and genetic evidence support the base-pairing of

---

FIGURE 6.9. A model for the pathway of spliceosome assembly. The diagram shows a simplified scheme for spliceosome assembly, incorporating observations from both yeast and vertebrate splicing systems. Substrate RNAs (the exons are depicted as boxes, the intron is shown as a line) with appropriate splice sites can assemble into an active splicing complex, termed the spliceosome (see text for references). An early step is recognition and binding of the 5'SJ by U1 snRNP. That complex adds U2 snRNP in a factor-dependent (U2AF) process, leading to the protection of the branch site. A second factor is shown helping U5 snRNP associate near the 3'SJ at about the same time that a complex snRNP containing U6 and U4 is added. Stabilization of the complex probably involves binding contacts between several pairs of snRNPs and factors (not diagrammed) as well as sequence recognition of the substrate sites. Departure of U4 from the complex is shown prior to the first transesterification (STEP 1). Additional components may join and depart from the spliceosome complex prior to the second transesterification (STEP 2). The spliced exon product is destined for export to the cytoplasm, the intron complex with the snRNPs is dissociated into subassemblies that presumably recycle, while the intron lariat RNA is turned over. This diagram principally follows the reviews by Padgett et al. (1986), Green (1986), and Guthrie and Patterson (1988), as well as the recent papers by Cheng and Abelson (1987) and Reed, Griffith, and Maniatis (1988).

U1-snRNP to the 5' splice site as one mode of binding and sequence recognition. Similar evidence supports base-pairing between U2-snRNP and the branch point sequence, the highly conserved UACUAACA motif in yeast. Other factors have been implicated in binding near (and presumably recognizing) the 3' splice junction. There is also good evidence that similarly sized complexes are formed at splice junctions in vivo (Osheim and Beyer, 1988). It seems clear that the process of splicing is specified by spliceosome assembly and that the components of the spliceosome are essential to splicing. There remains much to determine about the mechanism.

It is important for most introns in a particular organism to be spliced by a single set of accessory factors. However, some genes with multiple introns are spliced differently under certain circumstances. This topic is discussed thoroughly by Nadal-Ginard, et al., in Chapter 7 of this volume and thus only a few points germane to later sections will be made here. Differential splicing often involves an intron with one or more "atypical" features. Choosing when or how to splice such an intron can be accomplished by the action of intron- or gene-specific positive or negative control factors. For example, in a specific tissue or developmental stage, when a "typical" splice is avoided in favor of an "atypical" one, the choice can result from a factor that either promotes the unusual event or blocks the normal one. While a given organism may have a number of genes that engage in differential splicing, it is clear that no single factor can control all of them. Thus, among nuclear pre-mRNA introns, there are probably some cases of intron-specific (or subgroup-specific) splicing factors.

Trans-splicing of nuclear pre-mRNA introns was first demonstrated to be possible by Konarska and Sharp (1985) and Solnick (1985). Their findings suggested that components of the spliceosome probably contact each other without being first drawn together by binding to a single continuous RNA chain. For a brief time, such *trans*-splicing was a mere laboratory curiosity. However, it was later found that probably all mRNAs of trypanosomes share a common leader sequence that is added to separate transcripts by *trans*-splicing (Murphy et al., 1986; Sutton and Boothroyd, 1986). Later, several groups discovered that, while many nuclear-encoded mRNAs of *C. elegans* result from the usual sort of *cis*-splicing, some actin genes have a common leader sequence that results from *trans*-splicing (Krause and Hirsh, 1987). In both systems it has been shown that the leader RNA assembles as a special snRNP, termed SL SnRNP (Bruzik et al., 1988; Thomas et al., 1988; van Doren and Hirsh, 1988). In trypanosomes, this snRNP appears to exist in place of the U1-snRNP (Tschundi et al., 1986), while in *C. elegans* it co-exists with U1-snRNP (Thomas et al., 1988). Clearly the spliceosome can accommodate variations on the basic splicing theme.

Pre-tRNA Splicing

The pathway of splicing of pre-tRNAs is relatively straightforward, although it involves some very interesting activities and intermediates. It is clearly an enzyme-mediated, ordered set of reactions. The steps have been dissociated and the enzymes needed for each have been identified in several systems. Since most of the work in the field of pre-tRNA splicing has involved studies of yeast extracts and cloned pre-tRNA genes from yeast, most of this section will review these studies. The remainder will be an update on advances using extracts of an archaebacterium that allow several interesting comparisons to be made.

Transcripts of yeast genes can be accurately spliced in vitro with activities found in yeast (Pebbles et al., 1983), amphibian (Attardi et al., 1985), or mammalian cells (Filipowicz and Shatkin, 1983). This shows that such introns are universally recognized in eukaryotes and processed correctly by enzymes that are widely distributed. Presumably, pre-tRNAs with introns are also widely distributed, even in organisms for which tRNA genes with introns have yet to be described.

The pathway of tRNA intron splicing is summarized in Fig. 6.10. The intron is released as a linear product following two precise cleavages of the precursor by an endonuclease activity. This cleavage leaves the phosphates of the cleaved bonds either in the form of a $2'-3'$ cyclic phosphate or as a $2'$-phosphate (Peebles et al., 1983). The half-molecules remain associated via the base-pairing of the anticodon- and acceptor-stems and are the substrate for an RNA ligase activity. In yeast, this enzyme has multiple activities (Phizicky et al., 1986) including (1) cyclic phosphate hydrolase to yield the $2'$ phosphate; (2) polynucleotide kinase, utilizing the gamma phosphate of ATP to phosphorylate the $5'$ hydroxyl of the $3'$ half molecule; and (3) RNA ligase, which involves formation of a covalent adenylate–protein adduct, transfer of the adenylate to the $5'$ phosphate, and joining of the activated $5'$ end to the $3'$-hydroxyl with elimination of the adenylate moiety. In vitro, this junction product retains the $2'$-phosphate adjacent to the newly formed phosphodiester bond. There is an RNA ligase activity from wheat germ that carries out ligation with the tRNA half-molecule substrates, utilizing the same mechanism as that from yeast (Furneaux et al., 1983). Notably, mature tRNAs purified from cells lack detectable $2'$-phosphates at the splice junction position, so there must be a phosphatase that normally acts to remove them.

Recent evidence suggests that the RNA ligase forms a specific complex with the pre-tRNA. This complex then serves as the substrate for the endonuclease. In this way, the ligase gains exclusive access to the products of cleavage in vitro (Greer, 1987). Experiments using antibodies directed against the ligase show that this protein is located in nuclei and may be predominantly near the periphery (Clark and Abelson, 1987). The endonuclease from yeast has the properties of an integral membrane protein (Peebles et al., 1983). By implication it, too, is nuclear and it would make sense for the endonuclease to be embedded in the inner nuclear envelope. One interpretation of these results is that the ligase binds and shuttles the pre-tRNA to the endonuclease (Clark and Abelson, 1987). In this way, processing steps would be coordinated and coupled to tRNA export from the nucleus.

This attractive scenario may be specific to yeast, however. The endonuclease activity has also been isolated from amphibian cells, where it is nuclear in

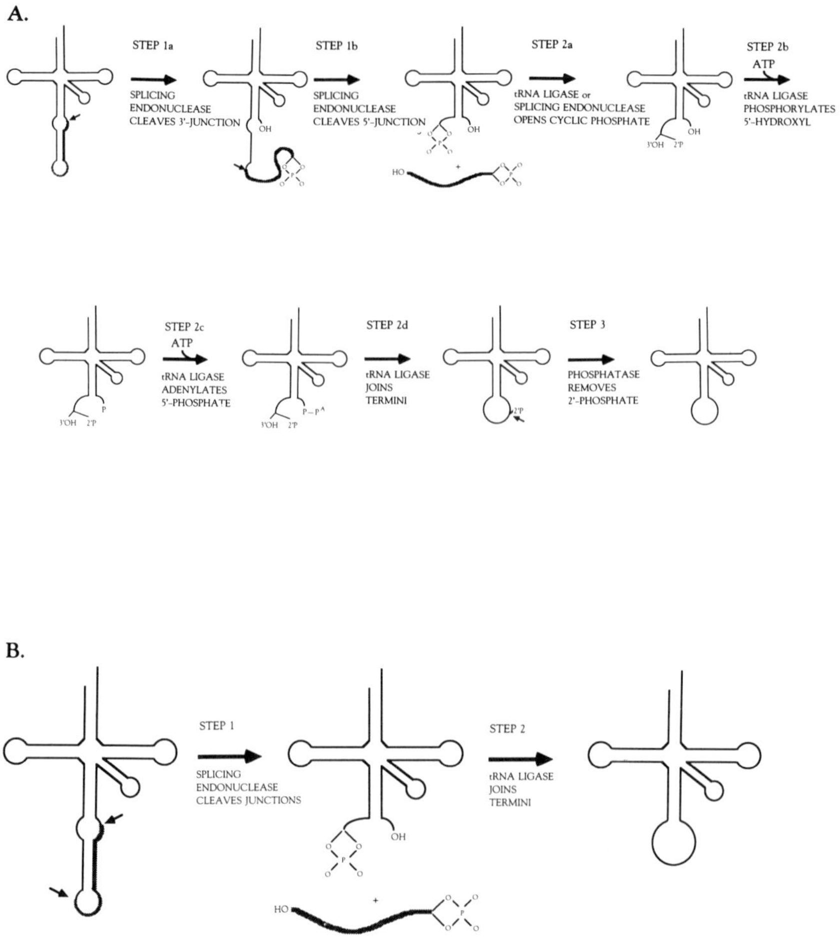

FIGURE 6.10. Pathways for pre-RNA splicing. Exons of the end-mature pre-tRNA are indicated by the thin lines, the intron is the thicker line. (A) The schematic diagram outlines the principal steps of the pathway operating in yeast cells. STEP1 is divided into two independent cleavage reactions at the splice junctions, but the particular order shown is not known to be general. Terminal hydroxyl (OH), monophosphate (P), and cyclic phosphate groups are indicated on the intermediates and products. (B) A distinct reaction pathway operates in animal cells. The key differences are due to the distinct mechanism of exon ligation.

location but evidently is not membrane bound (Attardi et al., 1985). This may reflect the different requirements and organization of the small and rapidly dividing yeast cell from the larger metazoan architecture. Yeast cells normally undergo nuclear division without dissolution of the nuclear envelope, in contrast to the usual process of mitosis in which the nuclear envelope is "disassembled" at every division.

Other differences also exist in the details of the ligation mechanism. In mammalian or amphibian cell extracts, the tRNA half-molecules are joined by an activity that converts the 2', 3' phosphodiester by an ATP-independent reaction (Filipowicz et al., 1983). The same extracts have an activity that can regenerate the cyclic phosphate from the monophosphate in an ATP-consuming reaction (Filipowicz et al., 1983). This pathway conserves the phosphate originally found at the 5' splice junction and does not generate the extra 2'-phosphate in the product. No additional ATP is needed to phosphorylate or activate the 5' end of the 3' half. Yet the products of cleavage by the amphibian endonuclease have the same structure and are interchangeable with those of the yeast endonuclease, in that ligase activity from either source can utilize half-molecules generated by either endonuclease.

Thompson and Daniels have recently developed an in vitro assay for an endonuclease specific for a pre-tRNA from the archaebacterium *Halobacterium volcanii* (Thompson and Daniels, 1988). It correctly excises the 104-nt intron of tRNA^{trp} from the same organism (and from the closely related *H. mediterranei*) yielding products with 5' hydroxyl and 2', 3' cyclic phosphate termini. It was somewhat unexpected that the extract was inactive with yeast pre-tRNA^{phe}, whose intron is located at the same site in the mature tRNA structure. Site-directed mutations of the archaebacterial tRNA substrate revealed that, unlike nuclear enzymes, major portions of the mature structure can be removed while retaining endonuclease activity at the correct sites. Similarly, deletions of intron sequences revealed that parts of the intron affect the efficiency but not the specificity of cleavage. It has been proposed that the substrate for the archaebacterial enzyme resembles an RNase III site. It will be interesting to see whether this apparent difference in activity is confirmed by further tests such as expression of the archaebacterial tRNA gene in yeast or in amphibian oocytes.

For tRNA splicing (excluding group I and group II introns of chloroplast tRNA genes) the role of trans-acting factors is best characterized. It is clear that two enzymes, an endonuclease and a ligase, are required for splicing and that they can act independently. Recent genetic studies in yeast, however, indicate that there are more genes involved in tRNA intron splicing than there are subunits for the two known enzymes (Winey and Culbertson, 1987; Peebles and O'Connor, unpublished data). Thus, it is likely that only a portion of the story is known.

Group I Splicing

Self-splicing was discovered by Tom Cech and his coworkers at the University of Colorado. They studied the intron of the large rRNA gene of the macronucleus of *Tetrahymena thermophila* and provided compelling evidence that its splicing could be entirely RNA-mediated (Kruger et al., 1982). Their initial studies were reported without their being aware that the rRNA intron is but one example of a large group of introns, most of which are found in organelle

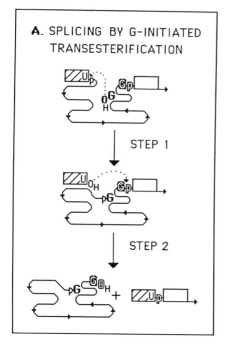
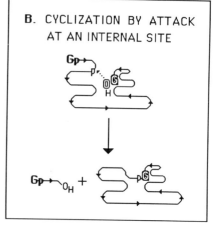
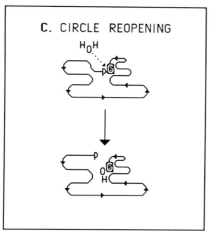
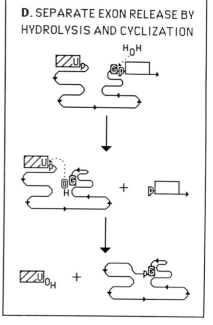

FIGURE 6.11. Splicing of group I introns and selected splicing-related reactions. The pre-RNA is diagrammed to indicate E1 (lined box) ending with U, intron (curved line to suggest the folded structure) ending in G, and E2 (open box), as well as the phosphodiesters (P) at the splice junctions. Bold, outline, and plain characters are used to distinguish the origin of particular groups as the reactions proceed to products. Arrowheads indicate the 5' to 3' direction. (A) Splicing by G-initiated transesterification. The guanosine cosubstrate (bold G_{OH}) is bound by the intron and provides the 3'-OH to attack the 5'SJ. The 3'OH of E1 attacks the 3' SJ, joining the exons and releasing the linear intron RNA with the extra G at the 5'end. (B) Cyclization by attack at an internal site. The 3'OH (outline) can attack an internal site

DNAs of fungi and plants. As noted above, the connection between the self-splicing intron of *Tetrahymena* and group I introns was made in 1983. Shortly thereafter, other workers extended Cech's findings, demonstrating that the first intron of the *cob* gene of *N. crassa* mtDNA (Garriga and Lambowitz, 1984), the rRNA intron of yeast mtDNA (and more recently introns 3 and 5a of the *coxI* gene and intron 5 of the *cob* gene), were shown to self-splice under similar conditions (van der Horst and Tabak, 1985; Gampel and Tzagoloff, 1987; Tabak et al., 1987; Partono and Lewin, 1988). Recently, group I introns of bacteriophage T4 (Chu et al., 1986; Gott et al., 1987) and intron 3 of the large rRNA gene of nuclear DNA of *Physarum* (Muscarella and Vogt, 1989) were also shown to self-splice.

The self-splicing of Group I introns is mediated by the folded structure of the intron RNA itself and proceeds through two ordered, transesterification steps (Fig. 6.11) (reviewed by Cech and Bass, 1986; Cech, 1988). The first involves guanosine or a guanosine nucleotide as the attacking nucleophile, which releases as intermediates the 5' exon with a 3' hydroxyl, and a linear IVS-3' exon with the G added to the intron 5' end. The second step is attack by the 3' hydroxyl of the first exon at the 3' splice junction to form the spliced exons and release the intron as a linear product with the extra G still at the 5' end. The phosphate at the splice junction is derived from the first nucleotide of the 3' exon. The G cosubstrate is attached via a phosphate derived from the first nucleotide of the intron. These reactions require moderate concentrations of monovalent cations, a divalent cation, usually magnesium, and the G cosubstrate. The *Tetrahymena* rRNA intron has been the most intensively analyzed, but all other self-splicing group I introns use the same basic pathway under similar reaction conditions.

It is very clear that the last 3–7 nucleotides of the 5' exon are essential in defining the 5' splice site. Their pairing with the internal guide sequence is amply supported by studies of in vitro mutants of the *Tetrahymena* intron (Inoue et al., 1985; Been and Cech, 1986; Waring et al., 1986). In vivo mutants of some yeast mitochondrial introns also reveal the same interaction (cited previously). This dependence on flanking coding sequences has no precedent with nuclear pre-mRNA or pre-tRNA introns.

The 3' splice junction of the *Tetrahymena* intron is very labile in vitro. When incubated at moderately high pH, it is cleaved by hydroxide ion. This reaction does not require GTP or even prior reaction at the 5' boundary (Inoue et al., 1987). Some of the fungal introns (specifically, the rRNA intron and aI3 of

---

←

to release a 5'-terminal oligonucleotide and form a shortened, circular intron. (C) Circle reopening. Circular intron products are subject to site-specific hydrolysis to produce shortened linear intron RNAs. (D) Separate exon release by hydrolysis and cyclization. In the absence of a suitable G-cosubstrate, pre-RNA is subject to site-specific hydrolysis at the 3'SJ, releasing E1:intron and E2 RNAs. The 3'OH of the E1:intron RNA can attack the 5'SJ to release E1 and full-length circular intron products.

yeast and *cob* intron 1 of *Neurospora*) carry out a reaction at the 3' splice junction that involves GTP attack there (van der Horst and Tabak, 1987; Garriga and Lambowitz, unpublished). Intron 3 of the yeast *coxI* gene also has an internal site for GTP attack that resembles the 5' boundary (Tabak et al., 1987). Both types of GTP attack at sites other than the 5' boundary divert molecules into pathways that do not lead to splicing. So far, there is no evidence for such reactions in vivo.

The excised *Tetrahymena* intron can carry out additional reactions (summarized in Fig. 6.11 and reviewed in Cech and Bass, 1986). These include cyclization of the intron with release of an oligonucleotide from the 5' end that contains the added G (Kruger et al., 1982). Circle formation has been confirmed in vivo for the *Tetrahymena* intron (cited in Cech et al., 1981) and for several introns of fungal mitochondria (Tabak et al., 1984; Dib-Hajj, Hoffman and Perlman, unpublished observations). The cyclized intron can reopen itself at the cyclization junction (Zaug et al., 1984). This shortened linear intron may cyclize at another internal site, and again reopen. It was first hypothesized that these reactions mimic the first step of self-splicing and so require formation of an analog of the P1 structure using the internal guide sequence and an upstream sequence resembling the end of the 5' exon (e.g., Inoue et al,. 1986). Clearly, an intron sequence is essential for circle formation; but recent in vitro mutants from Cech's group show that circle formation can be blocked by intron mutations that have no effect on 5' exon release by GTP attack or on splicing (Been and Cech, 1987). Thus, some analog of P1 is formed, but it does not use the same intron sequence.

The linear intron or a truncated transcript containing the catalytic core can also act on external RNAs to promote sequence-specific cleavage with G-addition, nucleotidyl transfer, and phosphotransfer. All of these reactions involve transesterification by guanosine nucleotide, either monomeric, or as the 3' end of the intron molecule. The intron has a binding site for guanosine. It also uses a binding site for RNA, including or overlapping the IGS, which determines the site-specificity by the rules of Watson-Crick base-pairing, to select its substrates. Alteration of the IGS changes the sequence selectivity of the intron catalyst. Analogs of G can also be introduced. These are used in many cases, but with altered binding affinity. The 5' splice site of Group I introns is always preceded by a U residue, while the 3' splice site is always after a G. The first step of splicing always requires a G. This suggests that recognition of G is important in both steps. In one representation, the first and second steps of the group I splicing reaction can be represented as the forward and reverse of the simple transesterification reaction:

$$X\text{p}U\text{p}N + G \xrightarrow{\text{IVS}} G\text{p}N + X\text{p}U$$

(where X and N stand for A, C, G or U (Kay and Inoue, 1987).

Proteins Assist Self-Splicing in Vivo

Genetic studies in several fungi have revealed genes encoding proteins needed for splicing individual or sets of group I introns. The most unexpected finding was that some intron reading frames in yeast encode a protein needed for efficient splicing of the intron itself (e.g., Lazowska et al., 1980). This intron-specific splicing factor (and others found since) was given the name *maturase*. The subsequent recognition that maturase-requiring introns resemble known self-splicing introns seemed paradoxical. It is generally agreed that maturases probably participate in splicing in vivo by binding to sites on the pre-mRNA and folding the RNA in preparation for splicing. However, at present only indirect evidence supports this hypothesis (e.g, Delahodde et al., 1985). It is an attractive one, though, since it is clear that the RNA contains all information needed for carrying out the steps of splicing. Interestingly, the three group I introns that require maturases in vivo do not appear to self-splice in vitro.

Each of the three known group I maturases is encoded by the intron that it helps splice. Two of them are intron-specific and are involved in the splicing of only one intron. However, the maturase encoded by bI4 assists the splicing of bI4 and aI4. Since aI4 requires a maturase but does not encode one, it follows that a maturase could even be encoded by a nuclear gene.

There is one clear example of a nuclear-encoded, intron-specific splicing factor: the *CBP2* gene of yeast (McGraw and Tzagoloff, 1983). This gene was isolated from a collection of nuclear, respiration-deficient mutants (so called, pet⁻ mutants) and was found to block the expression of the mitochondrial *cob* gene. Detailed analysis of the phenotype of *cbp2* mutants (and null alleles constructed by reverse genetics) revealed that the *CBP2* gene product is needed only for the splicing of *cob* I5, a group I intron that has no reading frame. Strains completely lacking *cob* I5 have no growth defect on respiratory carbon sources even when the *CBP2* gene has been deleted (Hill et al., 1985). Interestingly, bI5 self-splices in vitro (Gampel and Tzagoloff, 1987; Partono and Lewin, 1988) even though it clearly requires a protein for its splicing in vivo.

A number of other pet⁻ mutants also affect mitochondrial intron splicing. One gene, *MRS1*, appears dedicated to aiding the splicing of bI3, an intron that encodes a specific maturase (Kreike et al., 1986; 1987). Another gene, *MSS18*, is involved in the splicing of aI5b, although detailed studies of it indicate that the effect may be indirect (Seraphin et al., 1988). Other mutants are less specific in their inhibition of splicing (cf., Pillar et al., 1983; Dieckmann et al., 1982). A subset of these mutants have turned out to be leaky alleles of nuclear genes encoding mitochondrial tRNA synthetases (Myers et al., 1985; Pape et al., 1985). In several cases, the splicing defect appears to be an indirect consequence of the partial translation defect: the cells are deficient for some, but not all, maturases and different mutations affect different sets of introns.

The first clear example of a splicing factor needed for a group of related introns is the *cyt*-18 gene of *Neurospora crassa*. *Cyt*-18 mutants were chosen from a set of temperature-sensitive, cytochrome deficient mutants (Mannella et al., 1979). In initial studies of this mutant it was found that splicing of the group I intron of the large rRNA gene of mtDNA is blocked conditionally (Bertrand et al., 1982). Later, it was found that the splicing of most, but not all, group I introns is also blocked at the restrictive temperature (Collins and Lambowitz, 1985). Akins and Lambowitz (1987) cloned the *cyt*-18 gene and deduced from sequence analysis that it encodes the mitochondrial tyrosine-specific tRNA synthetase.

In earlier work, Lambowitz's group developed a protein-dependent in vitro system for splicing the mitochondrial rRNA gene intron that does not self-splice (Garriga and Lambowitz, 1986). They showed that extracts of *cyt*-18 mutants are inactive at the restrictive temperature. More recent studies clearly show that the *cyt*-18 protein plays a direct role in splicing (Majumder et al., 1989). The protein has been purified to virtual homogeneity and shown to promote the splicing of the rRNA gene intron. These data demonstrate that, at least in *Neurospora*, a single protein can participate in the splicing of a set of group I introns. Remarkably, this protein has another important function in mitochondria, namely, it is the essential tyrosyl-aminoacyl tRNA synthetase.

A similar situation may exist in baker's yeast. Dujardin et al. (1980) reported dominant nuclear suppressors of a *cobI* 4 maturaseless mutant that define the *nam2* gene (for *n*uclear *a*ccommodation of *m*itochondrial defect). Genetic studies showed that mutant alleles of the *nam2* gene require the reading frame of *coxI* I4 to be intact (Dujardin et al., 1983; Anziano et al., in preparation). Thus *nam2* alleles appear to activate an otherwise silent maturase (Labouesse et al., 1987). Cloning and sequencing of the gene (Labouesse et al., 1985) revealed that it encodes a mitochondrial leucine tRNA synthetase (Herbert et al., 1988). While there is as yet no direct proof that the *nam2* protein plays a direct role in splicing, it is intriguing that it is associated with a pair of introns (*cob* I4 and *coxI* I4) that require a maturase. Some additional data suggest that the *nam2* gene is needed for splicing other mitochondrial introns (Herbert et al., 1988). In that case, a group-specific splicing factor may interact with intron-specific ones to form splicing complexes that may be distinct for each intron.

Group II Introns

Our interest in the possibility that group II introns of yeast mtDNA might self-splice was based on Cech's initial studies of the *Tetrahymena* rRNA gene intron and the earlier report by Grivell's laboratory that following excision, group II introns accumulate in vivo as "circular" RNAs (Arnberg et al., 1980). Once it was shown that the excised *Tetrahymena* intron can form circles both in vitro and in vivo, we became intrigued with the possibility that a substantially different kind of intron might carry out similar reactions. Our initial studies of the self-splicing of intron 5g of the *coxI* gene (aI5g) (Peebles et al., 1986), together with those of van der Veen et al. (1986), demonstrated self-splicing and yielded the surprising finding that the excised intron RNA is a lariat rather than a circle.

Like group I introns, self-splicing group II introns undergo a pair of transesterification reactions. The initiating nucleophile in this case is the 2′ hydroxyl of an adenosine residue within domain 6 of the intron RNA. The first step is attack by this hydroxyl on the 5' splice junction that releases the 5' exon with a 3' hydroxyl end and a lariat IVS-3' exon intermediate. The released 5' exon attacks the 3' splice junction to produce the spliced exons plus the excised intron lariat. The phosphate at the splice junction is derived from the first nucleotide of the second exon, while the 2'–5' phosphodiester is from the first nucleotide of the intron. No net change in the number of phosphodiester linkages results, and thus there is no need for a nucleotide cosubstrate or high energy cofactor. This reaction requires divalent cations (magnesium) but neither ATP nor GTP as cosubstrates. Remarkably, the self-splicing of yeast aI5g requires the nonphysiological temperature of 45°C.

Initial surveys of group II intron sequences did not detect an intron sequence equivalent to the internal guide sequence of group I introns. However, elegant studies by Jacquier and Michel (1987) defined two sequences located in intron domain 1 of yeast aI5g that base pair with adjacent sequences near the 3' end of the 5' exon. The intron sequences are called EBS1 and EBS2 (for exon binding site) and the exon sequences are called IBS1 and IBS2 (for intron binding site). A survey of the sequences of other introns of the group revealed that most can form EBS1-IBS1 and EBS2-IBS2 pairings. At present, however, their findings have not been extended to another intron or supported by in vivo data.

In group I introns, self-splicing is not limited to just one intron of the group. Schmelzer and Schweyen (1986) showed that another intron of the IIB subgroup, intron 1 of the cytochrome *b* gene (bI1) of yeast, also self-splices under virtually identical conditions to those reported earlier for aI5g. Recently, Hebbar and Perlman (submitted) found that a group IIA intron of yeast mtDNA, intron 1 of the *coxI* gene (aI1), also self-splices. The self-splicing of aI1 is especially interesting because it is the only one of the three that encodes a maturase required for splicing in vivo (Carignani et al., 1983). Unlike group I introns, however, no group II intron from any other source tested through the summer of 1988 self-splices under any condition suitable for the yeast introns.

Group II introns mediate some other reactions (summarized in Fig. 6.12). They probably reflect the general catalytic activities of these RNAs. Under appropriate reaction conditions, group II intron-containing pre-mRNAs can cleave the 5' splice site by hydrolysis (rather than transesterification) (Jacquier and Rosbash, 1986; Jarrell et al., 1988b). A similar reaction is cleavage of transcript or exogenous RNA at sites resembling the 5' splice site (Jarrell et al., 1988b). Yeast aI5g and bI1 (but not aI1) can also perform a reaction termed *spliced exon reopening* in which the excised intron promotes the hydrolysis of the ligation site in spliced exon product (Jarrell et al., 1988b; R. Dietrich, unpublished). This is a seemingly futile process, but may provide clues to the nature of the actual catalytic mechanism. Our current hypothesis is that this reaction is related to the reversal of the second step of splicing (by hydrolysis rather than transesterification). Spliced exon reopening is unique to group II introns—group I introns do not have any analogous activity. Under some conditions, group II introns can form branches to an abnormal acceptor (Schmelzer

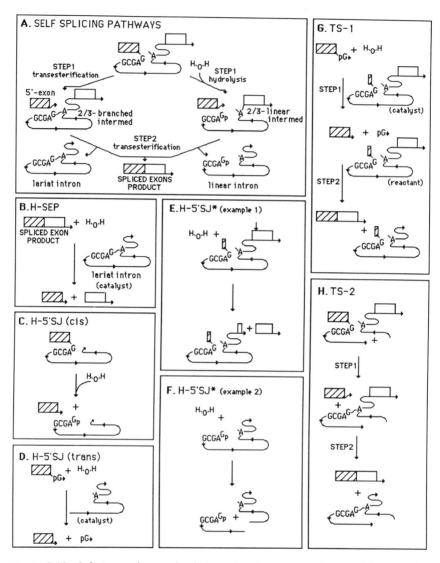

FIGURE 6.12. Splicing and several splicing-related reactions of group II introns. The pre-RNA is diagrammed to indicate E1 (lined box), intron (curved line to suggest the folded structure necessary for splicing) with conserved 5' terminal and branch A residues, and E2 (open box). Terminal phosphates (P) are indicated for intermediates and products. Arrowheads indicate the 5' to 3' direction. (A) Self-splicing pathways. Splicing can be initiated either by attack of the 2'OH of the branch A at the 5'SJ (transesterification) to produce a branched intermediate or by hydrolysis to yield a linear intermediate. The 3'OH of E1 attacks the 5'SJ to yield spliced exons and the excised intron product. (B) Hydrolysis of the spliced exon product (H-SEP). The intron catalyzes site-specific hydrolysis of the spliced product to yield separated E1 and E2 RNAs. (C) Incomplete pre-RNAs can carry out site-specific hydrolysis at the 5'SJ (H-5'SJ cis) to yield E1 and intron fragment RNAs. (D) Intron RNA

and Muller, 1987) or use an abnormal 5' end for branch formation (Hebbar, 1989).

In another sort of experiment, Jacquier and Rosbash (1986) showed that aI5g could use exogenous RNA as the 5' exon in the second step of splicing, a reaction they called "*trans*-splicing". Recently, several groups have reported that some group II introns of chloroplast DNA are interrupted by unrelated coding sequences so that the genes containing them are expressed via *trans*-splicing between separate transcripts (Koller et al., 1987; Zaita et al., 1987; Choquet et al., 1988). This type of *trans*-splicing has been modeled by Jarrell et al., (1988a) who divided yeast aI5g in domain 4. The resulting half-molecules were found to associate productively in a true *trans*-splicing reaction, carrying out branching and splicing to yield a Y-shaped intron containing the branch and the usual spliced exon product. The Y-shaped intron RNA retains the ability to carry out post-splicing reactions such as spliced-exon reopening and cleavage at internal sites resembling the 5' splice junction.

Jarrell et al. (1988a) further dissected the half-molecules to show that reformation of the interrupted domain 4 secondary structure (based on existing computer-generated models) is not responsible for the *trans* reactions. They even deleted all of domain 4 both in *cis* and in *trans* with retention of self-splicing activity. Instead, they found that domain 5, the most highly conserved internal sequence of group II introns, is essential for the reconstitution of the first reaction step from separate RNA molecules. In fact, they showed that the first step absolutely requires domain 5, while under some reaction conditions, it does not need domain 6, the site of the usual nucleophile of step one. These data reveal another novel feature of group II introns relative to group I: partial reactions can be carried out by RNAs lacking an element of the core structure.

The release of the intron as a lariat suggests a close relationship between this mechanism and that for nuclear introns. Though the same kind of structure is formed, there are two important differences. First, nuclear introns do not have their branch sites as close to the 3' end of the intron as group II introns. Second, excised group II intron lariats are quite stable in vivo (cf., Arnberg et al., 1980) while excised nuclear introns are turned over efficiently (e.g., Zeitlin

---

catalyzes hydrolysis of transcripts containing the 5'SJ (H-5'SJ *trans*) to yield E1 and intron fragment RNAs. (E) Transcripts lacking a complete E1 carry out hydrolysis at a site in E2 that shares partial sequence similarity with the 5'SJ (H-5'SJ*) to yield truncated E2 and extended intron RNAs. (F) Transcripts or products lacking both E1 and E2 carry out H-5'SJ* at a site within D2 to yield two intron fragments. (G) Pairs of transcripts (half-molecules) carry out splicing in trans. The first class of *trans*-reactions (TS-1) involves E1 and intron:E2 transcripts that associate and proceed through splicing to yield spliced exons, a linear intron, and a small intron fragment. (H) Nonoverlapping half-molecules containing E1:intron-fragment or intron-fragment:E2 can associate in certain cases to carry out a second class of *trans*-splicing (TS-2). The products are spliced exons and branched intron RNAs. This class of *trans*-splicing occurs in the cases of split genes reported in some chloroplast systems.

and Efstradiatis, 1983; Domdey et al., 1984). Nuclei contain an activity that reopens 2'-5' branches that may play an important role in turnover of lariat RNAs (Ruskin and Green, 1985), while organelles do not appear to have an analogous activity.

Another similarity is also evident in the boundary sequences of group II introns—these resemble those of nuclear introns. Again the similarity has its limitations; it is clear that the conserved GU at the 5' ends of nuclear introns is essential for correct splicing in vivo and in vitro. However, not all group II introns have a G and U at the 5' boundary even though they splice in vivo. In addition, mutation of the conserved G of yeast aI5g has little effect on splicing in vitro (Hebbar, 1989). The similarity may extend to analogies between the domains of group II introns and the secondary structures formed by the binding of snRNAs to nuclear introns. Domain 6 displays the A residue that accepts the branch as an unpaired bulge in a pyrimidine-rich strand usually base paired with a purine-rich strand. This sort of structure has been proposed for the U2-snRNP bound to the branch site UACUAACA of yeast nuclear introns (Parker et al., 1987). Unlike nuclear introns, however, it is clear that splicing of group II introns can proceed in vitro without branch formation during the first step (Jacquier and Rosbash, 1986; Jarrell et al., 1988a, b). Recently, Hebbar (1989) has deleted the branch site containing domain 6 of aI5g with little effect on the rate or extent of in vitro splicing.

The group II transesterification mechanism is similar to that for group I introns—indeed, the second step is almost the same in terms of chemical reaction for group I, group II, and nuclear introns. Like group I introns, the folded RNA structure of a group II intron is crucial to reactivity. But, neither the boundary sequences nor the details of the folded structure of group I are evident in the group II introns. There are group II introns whose splicing is dependent on genetically defined *trans*-acting factors, presumably proteins (Carignani et al., 1983; Mecklenberg, 1986). The mechanism of the protein assistance is not understood at present; but all of the group II introns use this pathway in vivo, as evidenced by the accumulation of the lariat intron products.

GROUP I AND GROUP II INTRONS AS MOBILE GENETIC ELEMENTS

Unlike nuclear pre-mRNA and tRNA introns, which appear to be persistent components of genomic lineages, group I and group II introns shown clear signs of being mobile genetic elements. In this section we review some data regarding the groups of introns that pertain to this issue.

Introns as Optional Components of Genes

Group I and Group II introns are often optional sequences. Careful studies of sets of strains, races, or species of *Saccharomyces* (Ralph, 1987; Dib-Hajj et al., in preparation), *Schizosaccharomyces* (Trinkl and Wolf, 1986; Zimmer et al., 1987), *Podospora* (Cummings et al., 1986), *Neurospora* (Collins et al.,

Table 6.1. Intron Composition of the *CoxI* Gene of Yeast Mitochondrial DNA*

Species	Intron									
	aI1	aI2	aI3a	aI3b	aI3g	aI4a	aI4b	aI5a	aI5b	aI5g
S. cerevisiae										
strain D273	+	+	+	−	−	+	−	−	−	+
strain ID41-6/161	+	+	+	−	−	+	−	+	+	+
S. capensis	−	−	+	−	+	−	+	−	+	+
S. ellipsoideus	+	−	+	−	+	−	−	−	−	−
S. norbensis	+	+	+	−	+	−	−	−	+	+
S. coreanus	−	+	+	−	−	+	−	+	+	+
S. diastaticus	+	−	+	−	−	+	+	−	+	+
S. uvarum	−	+	+	−	−	−	−	−	−	+
S. douglasii	−	−	−	+	+	−	?	?	?	?

*Mitochondrial DNA was isolated from each of the *Saccharomyces* species shown and the presence of the seven known introns of that gene (c.f., Bonitz et al., 1980; Hensgens et al., 1983) was evaluated using intron-specific probes of DNA blots (see Ralph, 1986). The first two lines of the Table summarize the five- and seven-intron forms of the gene found in most standard laboratory strains of *S. cerevisiae*. The other species analyzed contain some of those introns in various combinations. Also, two new introns were identified in some organisms; these were sequenced and found to be group I introns with reading frames (Dib-Hajj et al., in preparation). Introns present in the *CoxI* gene of *S. douglasii* were described by Kotylak et al. (1985).

1983), *Tetrahymena* (Sogin et al., 1986), *Physarum* (Muscarella and Vogt, 1989), *Chlamydomonas* (Lemieux and Lee, 1987) and T-even bacteriophages (Quirk et al., 1989) reveal that almost any group I or group II intron found in one isolate may be absent from the same gene of another isolate. This is despite the finding that the organisms being compared are extremely closely related (where sequence data are available, usually over 95% identical) based on the primary sequence of exons.

Table 6.1 summarizes intron variation in the *coxI* gene in a set of *Saccharomyces* isolates (Kotylak et al., 1985; Ralph, 1987; Dib-Hajj et al., in preparation). Ten different introns are present in the gene, summing over the entire set of isolates. However, the largest number of introns in a single gene is seven and several different combinations of seven introns were observed. It is important that none of those introns is closely related to any other and that at a given intron location only one intron is present or absent. Seraphin et al. (1987) have constructed a *coxI* gene lacking all of the introns and reported that it is functional. Of course, this test may not be sufficient to conclude that introns are not involved in *coxI* gene expression in a more "natural" environment.

The *coxI* gene has been sequenced in mtDNA from other fungi. While it sometimes has introns, they are rarely found at the locations seen in yeast. Thus, either the progenitor of the fungal *coxI* gene had very many introns, or else the introns are relatively recent acquisitions. It is noteworthy that most mitochondrial genes are quite ancient—with homologs present in most types of bacteria. In metazoans, mitochondrial genes lack introns totally.

If the full set of group I introns in all organisms is surveyed, a few examples of closely related introns in different genes or different organisms are found. Such situations are consistent with a model that assumes movement of an intron from one site (or organism) to another followed by sequence divergence. The clearest example of closely related introns at different sites in the same genome are the group I introns bI4 and aI4 of yeast mtDNA (Bonitz et al., 1980; Anziano et al., 1982). They are about 70% identical in nucleotide sequence and have long reading frames (see above) that are similarly related at the amino acid sequence level. Likewise, an intron closely resembling the rRNA intron of yeast mtDNA is present in the ATPase subunit 9 gene of a *Kluyveromyces* isolate (Dujon et al., 1986).

There are some examples of closely related introns at the same site in different organisms. For example, bI3 of yeast and bI1 of *Aspergillus* differ somewhat in length due to some large insertions and/or deletions but otherwise have a high level of sequence homology (Holl et al., 1985). Lang (1984) reported that intron 2 of the *coxI* gene of *S. pombe* mtDNA closely resembles intron 3 of that gene in *A. nidulans*. An intron closely related to the rRNA gene intron of yeast mtDNA is present in the rRNA gene of mtDNA of *K. thermotolerans* (Jacquier and Dujon, 1983). And finally, intron 3 of the nuclear rRNA gene of *Physarum* closely resembles the rRNA gene intron of *Tetrahymena* (Muscarella and Vogt, 1989).

Group II introns of fungal mtDNAs are highly variable when different organisms are compared. However, introns of plant organelle genes are much more highly conserved between species. For example, the primary sequences of the *coxII* gene of mtDNA of corn, wheat, and rice are over 98% identical; and each gene has a single group II intron at the same site (Fox and Leaver 1981; Bonen et al., 1984; Kao et al, 1984). These introns are nearly identical in sequence but differ in length due to insertion and/or deletion. Because the insertion in the rice gene is flanked by a repeated sequence, it has been suggested that the intron may have "captured" a mobile element (Xiong and Eickenbush, 1988). Interestingly, the extra sequence is located within intron domain 4, a site where it is likely that sequences can be added or deleted with little consequence on splicing (see Jarrell et al., 1988b). Similarly, where a group II intron is present in cpDNA of one plant, a nearly identical intron is usually present at the same site in other closely related plants.

An interesting situation exists in yeast mtDNA that may resemble the variation in *coxII* gene introns of plant mtDNA. The mitochondrial genome of yeast is punctuated by a large family of G+C-rich sequences (de Zamoraczy and Bernardi, 1985). These sequences are highly variable when mtDNAs of different strains are compared and clear evidence exists that at least one family of them is mobile (reviewed by Butow et al., 1985). Most G+C-rich clusters are located outside of coding regions and few are present in introns; however, there are now a number of examples of G+C-rich clusters that are optional features of group I and group II introns; (Weiler, 1987; Hoffman, 1989; R. C. Dietrich, unpublished). In each case, the cluster is inserted within a region of the intron that is not thought to be involved in forming the core structure. Perhaps related to this is the report of an intron of the ND1 gene of *N. crassa* mtDNA that is conserved between species but has a different reading frame inserted at a different site in the intron of mtDNA of *N. intermedia* (Mota and Collins, 1988).

Lastly, Kjems and Garrett (1988) have reported an intriguing situation in an archaebacterium where a tRNA-like intron may have captured a mobile group I intron. They were studying the mechanism of splicing of the only known group I-like intron in archaebacteria, the intron of the large rRNA gene of *D. mobilis*. This intron does not self-splice and was found to depend on a protein-containing extract of the organism for splicing in vitro. However, that reaction does not require GTP and yields reaction intermediates and products that are quite typical for the tRNA intron splicing pathway rather than for the group I intron pathway. Modeling indicates that the exon sequence flanking the intron is rather tRNA-like.

These results are quite striking for several reasons. First, they define the first example of an rRNA intron that appears to use the tRNA intron splicing pathway. Second, the intron is very long and contains an open reading frame, typical group I intron structures, and conserved sequences. It is possible that the intron is really a tRNA intron that has captured a mobile group I intron. Altogether, the examples summarized in this section provide strong circumstantial evidence that group I and group II introns may be mobile elements and that they interact with other intron-related or non-intronic mobile sequences.

Group I Intron Mobility

Several lines of evidence reveal extant processes that may account for intron acquisition in relatively recent evolutionary time. First, some group I introns are actively mobile in present-day organisms. This was first documented for the rRNA intron of yeast mtDNA (Dujon, 1980). In crosses between functionally equivalent wild-type strains that either have or lack the optional intron, virtually all recovered progeny contain the intron, even though genetic markers at distant sites on the parental genomes are transmitted coordinately (reviewed by Dujon, 1981; Butow, 1985).

The broad outline of the mechanism of this form of intron mobility or "infectivity" is now known. The intron has a reading frame that encodes a sequence-specific DNA endonuclease (Jacquier and Dujon, 1985; Macreadie et al., 1985; Colleaux et al., 1986). It cleaves intronless alleles in mated cells (where the two forms of the gene share a common cytoplasm) very near the site of intron insertion (Zinn and Butow, 1985). The cleavage site is absent from the intron-containing parent because the recognition sequence is interrupted by the intron. Following double-strand cleavage, a form of recombination ensues that closely resembles the better understood mating-type switching in the yeast nucleus (cf., Strathern et al., 1982).

It is now clear that this phenomenon, of obvious survival value to the intron, is not limited to this one example. Genetic studies indicate that an intron of the rRNA gene of *Chlamydomonas,* and several optional sequences of the mtDNA of *Neurospora,* engage in similar mobility (Mannella and Lambowitz, 1979; Lemieux and Lee, 1987). A second group I intron of yeast mtDNA, *coxI* intron 4a, has now been shown to encode an endonuclease with different specificity that promotes its preferential transmission (Wenzlau et al., 1989; Delahodde et al., 1989). Surprisingly, two of the group I introns of bacteriophage T4 (Quirk et al., 1989) and an intron of the rRNA gene of nuclear DNA of *Physarum* (Muscarella and Vogt, 1989) also exhibit similar mobility in appropriate crosses. Thus, as recently as 1989, an apparently obscure phenomenon (a mobile group I intron) has been associated with a significant fraction of group I introns and is no longer limited to the mitochondrion of yeast. However, it should be added that clear evidence exists that some group I introns of yeast mtDNA do not show mobility when tested analogously.

In all of these examples of mobile introns, the intron can encode a protein and good evidence exists that the protein is a sequence-specific endonuclease involved in the mobility process. One of them, the omega endonuclease of the rRNA gene intron of yeast mtDNA, is known to play no role in splicing the intron. It is very interesting that the aI4a encoded endonuclease is closely related in primary sequence to the bI4 encoded maturase and that, under certain circumstances, the aI4a protein can substitute for the bI4 maturase (summarized by Wenzlau et al., 1989). At present it is not known whether these two activities share certain domains of the protein; however, it is clear that both functions can contribute to the survival of the intron, though in very different ways. Since group I introns are fundamentally self-splicing, Wenzlau et al. have suggested that the endonuclease function may be the more ancient one.

Reverse Transcription and Group II Introns

So far, no group II intron has been found to engage in intron conversion. It is interesting that while few group II introns have ORFs, the proteins encoded by most of them have a striking similarity in amino acid sequence to retroviral reverse transcriptase (Michel and Lang, 1985; Xiong and Eickbush, 1988). It may be relevant that a group II intron of *Podospora* can be cleanly excised from mtDNA to yield a circular DNA copy of the intron (Osiewacz and Esser, 1984). In yeast mitochondria, intron excision events (resulting in intron loss but no other discernable consequence) are known (e.g., Gargouri et al., 1983; Hill et al., 1985; Seraphin et al., 1987). Remarkably, events in which multiple adjacent introns are excised simultaneously, with retention of exon sequences, are not rare. It appears that reverse transcription followed by gene conversion is an attractive model for this phenomenon.

Recent findings in Lambowitz's laboratory make the connection between mitochondrial introns and reverse transcription more enticing. Some isolates of *Neurospora* have circular DNA plasmids in their mitochondria (Collins et al., 1981; Stohl et al., 1982). The closely-related Mauriceville and Varkud plasmids are abundantly transcribed to yield single, major full length RNA, even though they are not needed for mitochondrial function (Nargang et al., 1984; Akins et al., 1988). The primary sequence of the plasmid contains clear examples of group I intron-conserved sequences. The long open reading frame, however, has clear homology to reverse transcriptase (a feature of group II intron reading frames) although it lacks conserved sequences typical of group II introns (see also Michel and Lang, 1985). Akins et al. (1986) have characterized slow-growing mutants that appear to result from the insertion of plasmid DNA sequences into a gene of mtDNA. Kuiper and Lambowitz (1988) have now detected a remarkable reverse transcriptase activity in RNP particles from plasmid-containing strains of *Neurospora,* and they have provided convincing evidence that the major transcript of the plasmid is reverse-transcribed. Taken together, these data provide reason to conclude that retrotransposition can occur in mitochondria and that intron-related sequences can change position through that process.

REFERENCES

Abelson, J. (1979). Ann. Rev. Biochem. 48:1035.
Akins, R. A., Grant, D. M., Stohl, L. L., Bottorff, D. A., Nargang, F. E., and Lambowitz, A. M. (1988). J. Molec. Biol. 204:1.
Akins, R. A., Kelley, R. L., and Lambowitz, R. A. (1986). Cell 47:505.
Akins, R. A., and Lambowitz, A. M. (1987). Cell 50:331.
Alexander, N. J., Perlman, P. S., Hanson, D., and Mahler, H. R. (1980). Cell 20:199.
Andreadis, A., Hsu, Y.-P., Kohlhaw, G. B., and Schimmel, P. (1982). Cell 31:319.
Anziano, P. Q., Hanson, D. K., Mahler, H. R., and Perlman, P. S. (1982). Cell 30:925.
Arnberg, A. C., van Ommen, G. J. B., Grivell, L. A., van Bruggen, E. F. J., and Borst, P. (1980). Cell 19:313.
Attardi, D. G., Margarit, I., and Tocchini-Valentini, G. P. (1985). EMBO J. 4:3289.
Baldi, M. I., Mattoccia, E., and Tocchini-Valentini, G. P. (1983). Cell 35:109.
Banroques, J., Delahodde, A., and Jacq, C. (1986). Cell 46:837.
Barta, A., Sommergruber, K., Thompson, D., Hartmuth, K., Matzke, M. A., and Matzke, A. J. M. (1986). Plant Mol. Biol. 6:347.
Bass, B. L., and Cech, T. R. (1986). Biochemistry 25:4473.
Been, M. D., Barfod, E. T., Burke, J. M., Price, J. V., Tanner, N. K., Zaug, A. J., and Cech, T. R. (1987). Cold Spring Harbor Symp. Quant. Biol. 52:147.
Been, M. D., and Cech, T. R. (1985). Nucleic Acids Res. 13:3389.
Been, M. D., and Cech, T. R. (1986). Cell 47:207.
Been, M. D., and Cech, T. R. (1987). Cell 50:951.
Beggs, J. D., van den Berg, J., van Ooyen, A., and Weissman, C. (1980). Nature 283:835.
Berget, S. M., Moore, C., and Sharp, P. A (1977). Proc. Natl. Acad. Sci. USA 74:3171.
Bertrand, H., Bridge, P., Collins, R. A., Garriga, G., and Lambowitz, A. M. (1982). Cell 29:517.
Beyer, A. L., and Osheim, Y. N. (1988). Genes Dev. 2:754–765.
Black D. L., Chabot, B., and Steitz, J. A. (1985). Cell 42:737.
Blumenthal, T., and Thomas, J. (1988). Trends in Genetics 4:305.
Bonen, L., Boer, R. H., and Gray, M. W. (1984). EMBO J. 3:2531.
Bonitz, S. G., Coruzzi, G., Thalenfeld, B. E., Tzagoloff, A., and Macino, G. (1980). J. Biol. Chem. 255:11927.
Breathnach, R., Benoist, C., O'Hare, K., Gannon, F., and Chambon, P. (1978). Proc. Natl. Acad. Sci. USA 75:4853.
Breathnach, R., and Chambon, P. (1981). Ann. Rev. Biochem. 50:349.

Brody, E., and Abelson, J. (1985). Science 228:963.
Brown, J. W. S. (1986). Nucleic Acids Res. 14:9549.
Brown, J. W. S., Feix, G., and Frendewey, D. (1986). EMBO J. 5:2749.
Bruzik, J. P., van Doren, K., Hirsh, D., and Steitz, J. A. (1988). Nature, 335:559.
Burke, J. M., Belfort, M., Cech, T. R., Davies, R. W., Schweyen, R. J., Shub, D. A., Szostak, J. W., and Tabak, H. F. (1987). Nucleic Acids Res. 15:7217.
Burke, J. M., Irvine, K. D., Kaneko, J. J., Kerker B., Oettgen, A. B., Tierney, W., Williamson, C., Zaug, A. J., and Cech, T. R. (1986). Cell 45:167.
Burke, J. M., and RajBhandary, U. L. (1982). Cell 31:509.
Butow, R. A. (1985). Trends in Genetics 1:81.
Butow, R. A., Perlman, P. S., and Grossman, L. I. (1985). Science 228:1496.
Carignani, G., Groudinski, O., Frezza, D., Schiavon, E., Bergantino, E., and Slonimski, P. P. (1983). Cell 35:733.
Cech, T. R. (1988). Gene 73:259.
Cech, T. R., and Bass, B. L. (1986). Ann. Rev. Biochem. 55:599.
Cech, T. R., Tanner, N. K., Tinoco, I., Weir, B. R., Zuker, M., and Perlman, P. S. (1983). Proc. Natl. Acad. Sci. USA 80:3903.
Cech, T. R., Zaug, A. J., and Grabowski, P. J. (1981). Cell 27:487.
Chabot, B., and Steitz, J. A. (1987). Molec. Cell. Biol. 7:281.
Chen, C., Malone, T., Beckendorf, S. K., and Davis, R. L. (1987). Nature 329:721.
Cheng, S. C., and Abelson, J. (1987). *Genes Dev.* 1:1014.
Choffat, Y., Suter, B., Behra, R., and Kubil, E. (1988). Molec. Cell. Biol. 8:3332.
Choquet, Y., Goldschmidt-Clermont, M., Girard-Bascou, J., Kuck, U., Bennoun, P., and Rochaix, J.-D. (1988) Cell 52:903.
Chow, L. T., Gelinas, R. E., Broker, T. R., and Roberts, R. J. (1977). Cell 12:1.
Chu, F. K., Maley, G. F., Maley, F., and Belfort, M. (1984). Proc. Natl. Acad. Sci. USA 81:3049.
Chu, F. K., Maley, G. F., West, D. K., Belfort, M., and Maley, F. (1986). Cell 45:157.
Clark, M. W., and Abelson, J. (1987). J. Cell Biol. 105:1515.
Colleaux, L., d'Auriol, L., Betermier, M. M., Cottarel, G., Jacquier, A., Galibert, F., and Dujon, B. (1986). Cell 44:521.
Colleaux, L., D'Auriol, L., Galibert, F., and Dujon, B. (1988). Proc. Natl. Acad. Sci. USA 85:6022.
Collins, R. A., and Lambowitz, A. M. (1985). J. Molec. Biol. 184:413.
Collins, R. A., Stohl, L. L., Cole, M. D., and Lambowitz, A. M. (1981). Cell 24:443.
Conrad, H. M. (1987). Ph.D. thesis, The Ohio State University, Columbus.
Crick, F. (1979). Science 204:264.
Daniels, C. J., Gupta, R., and Doolittle, W. F. (1985). J. Biol. Chem. 260:3132.
Davies, R. W., Waring, R. B., Ray, J. A., Brown, T. A., and Scazzocchio, C. (1982). Nature 300:719.
Debuchy, R., and Brygoo, Y. (1985). EMBO J. 4:3553.
Delahodde, A., Banroques, J., Becam, A. M., Goguel, V., Perea, J., Schroeder, R., and Jacq, C. (1985). In Achievements and Perspectives of Mitochondrial Research, Volume 1: Biogenesis. E. Quagliariello, E. C. Slater, F. Palmieri, C. Saccone, and A. M. Kroon, eds. Elsevier Science Publishers, Amsterdam, p. 79.
Delahodde, A., Goguel, V., Becam, A. M., Creusot, F., Banroques J., and Jacq, C. (1989). Cell 56:431.
De la Salle, H., Jacq, C., and Slonimski, P. P. (1982). Cell 28:721.

Del Ray, F. J., Donohue, T., and Fink, G. R. (1982). Proc. Natl. Acad. Sci. USA 79:4138.
DeRobertis, E. M., Black, P., and Nishikura, K. (1981). Cell 23:89.
De Zamaroczy, M., and Bernardi, G. (1985). Gene 33:1.
Dieckmann, C. L., Pape, L. K., and Tzagoloff, A. (1982). Proc. Natl. Acad. Sci. USA 79:1805.
Domdey, H., Apostol, B., Lin, R.-J., Newman, A., Brody, E., and Abelson, J. (1984). Cell 39:611.
Dujardin, G., Labouesse, M., Netter, P., and Slonimski, P. P. (1983). In Mitochondria 1983, Schweyen, R. J., Wolf, K., and Kaudewitz, F., eds. W. de Gruyter, Berlin, p. 233.
Dujardin, G., Pajot, P., Groudinsky, O., and Slonimski, P. (1980). Mol. Gen. Genet. 179:469.
Dujon, B. (1980). Cell 20:185.
Dujon, B. (1981). In The Molecular Biology of the Yeast Saccharomyces: Life Cycle and Inheritance, Strathern, J., Jones, E., and Broach, J., eds. Cold Spring Harbor Laboratory, New York, p. 505.
Dujon, B., Colleaux, L., Jacquier, A., Michel, F., and Monteilhet, C. (1986). In Extrachromosomal Elements in Lower Eukaryotes, Wickner, R., Hinnebush, A., Gunsalus, I., Lambowitz, A., and Hollander, A. eds. Plenum Press, New York, p. 5.
Etchevery, T., Colby, D., and Guthrie, C. (1979). Cell 18:11.
Falkenthal, S., Parker, V. P., and Davidson, N. (1985). Proc. Natl. Acad. Sci. USA 82:449.
Filipowicz, W., Kanarska, M., Gross, H. J., and Shatkin, A. J. (1983). Nucleic Acids Res. 11:1405.
Filipowicz, W., and Shatkin, A. J. (1983). Cell 32:547.
Fox, T., and Leaver, C. (1981). Cell 26:315.
Fradin, A., Jove, R., Hemenway, C., Heiser, H. D., Manley, J. L., and Prives, C. (1984). Cell 37:927.
Frendeway, D., and Keller, W. (1985). Cell 42:355.
Furneaux, H., Pick, L., and Hurwitz, J. (1983). Proc. Natl. Acad. Sci. USA. 80:3933.
Gampel, A., and Tzagoloff, A. (1987). Molec. Cell. Biol. 7:2545.
Gamulin, V., Mao, J., Appel, B., Sumner-Smith, M., Yamao, F., and Soll, D. (1983). Nucleic Acids Res. 11:8537.
Gargouri, A., Lazowska, J., and Slonimski, P. (1983). In Mitochondria 1983, Schweyen, R., Wolf, K., and Kaudewitz, F., eds. De Gruyter, Berlin, p. 259.
Garriga G., and Lambowitz, A. M. (1984). Cell 39:631.
Garriga, G., and Lambowitz, A. M. (1986). Cell 46:669.
Gerke, V., and Steitz, J. A. (1986). Cell 47:973.
Goodall, G. J., and Filipowicz, W. (1988) Cell 58:473.
Goodman, H. M., Olson, M. V., and Hall, B. D. (1977). Proc. Natl. Acad. Sci. USA 74:5453.
Gott, J. M., Shub, D. A., and Belfort, M. (1986). Cell 47:81.
Gott, J. M., Zeeh, A., Bell-Pedersen, D., Ehrenman, K., Belfort, M., and Shub, D. (1989). Molec. Cell. Biol., in press.
Gouilloud, E., and Clarkson, S. G. (1986). J. Biol. Chem. 261:486.
Grabowski, P. J., Padgett, R. A., and Sharp, P. A. (1984). Cell 37:415.
Grabowski, P. J., Seiler, S. R., and Sharp, P. A. (1985). Cell 42:345.
Grabowski, P. J., and Sharp. P. A. (1986). Science 233:1294.

Green, M. R. (1986). Ann. Rev. Genet. 20:671.
Green, M. R., Grimm, M. F., Goewert, R. R., Collins, R. A., Cole, M. D., Lambowitz, A. M., Heckman, J. E., Yin, S., and RajBhandary, U. L. (1981). J. Biol. Chem. 256:2027.
Greer, C. L. (1987). Mol. Cell. Biol. 6:635.
Greer, C. L., Soll, D., and Willis, J. (1987). Molec. Cell. Biol. 7:76.
Guthrie, C., and Abelson, J. (1982). In Molecular Biology of the Yeast Saccharomyces: Metabolism and Gene Expression, Strathern, J. N., Jones, E. W., and Broach, J. R., eds. Cold Spring Harbor Laboratory, Cold Spring Harbor, New York, p. 487.
Guthrie, C., and Patterson, B. (1988). Ann. Rev. Genet. 22:387.
Haid, A., Schweyen, R. J., Bechmann, H., Kaudewitz, F., Solioz, M., and Schatz, G. (1979). Europ. J. Biochem. 94:451.
Halbreich, A., Pajot, P., Foucher, M., Grandchamp, C., and Slonimski, P. P. (1980). Cell 19:321.
Hall, D. H., Povinelli, C. M., Ehrenman, K., Pedersen-Lane, J., Chu, F., and Belfort, M. (1987). Cell 48:63.
Hanson, D. K., Lamb, M. R., Mahler, H. R., and Perlman, P. S. (1982). J. Biol. Chem. 257:3218.
Hanson, D. K., Miller, D., Mahler, H. R., Alexander, N. J., and Perlman, P. S. (1979). J. Biol. Chem. 254:2480.
Hartmuth, K., and Barta, A. (1986). Nucleic Acids Res. 14:7513.
Hebbar, S. K. (1989). Ph.D. Thesis, The Ohio State University, Columbus, Ohio.
Hensgens, L. A. M., Bonen, L., de Haan, M., van der Horst, G., and Grivell, L. A. (1983). Cell 32:379.
Herbert, C. J., Labouesse, M., Dujardin, G., and Slonimski, P. P. (1988). EMBO J. 7:473.
Hernandez, N., and Keller, W. (1983). Cell 35:89.
Hill, J., McGraw, P., and Tzagoloff, A. (1985). J. Biol. Chem. 260:3235.
Hoffman, P. (1989). Ph.D. Thesis, The Ohio State University, Columbus, Ohio.
Holl, J., Rodel, G., and Schweyen, R. J. (1985). EMBO J. 4:2081.
Hottinger, H., Pearson, D., Yamao, F., Gamulin, V., Cooley, L., Cooper, T., and Soll, D. (1982). Molec. Gen. Genet. 188, 219.
Huiet, L., Tyler, B. M., and Giles, N. H. (1984). Nucleic Acids Res. 12, 5757.
Inoue, T., Sullivan, F. X., and Cech, T. R. (1985). Cell 43:431.
Inoue, T., Sullivan, F. X., and Cech, T. R. (1986). J. Molec. Biol. 189:143.
Jacq, C., Pajot, P., Lazowska, J., Dujardin, G., Claisse, M., Groudinsky, O., de la Salle, H., Grandchamp, C., Labouesse, M., Gargouri, A., Guiard, B., Spyridakis, A., Dreyfus, M., and Slonimski, P. P. (1982). In Mitochondrial Genes, Slonimski, P., Borst, P., and Attardi, G., eds. Cold Spring Harbor Laboratory Press, Cold Spring Harbor, New York, p. 155.
Jacquier, A., and Dujon, B. (1983). Molec. Gen. Genet. 192:487.
Jacquier, A., and Dujon, B. (1985). Cell 41:383.
Jacquier, A., and Michel, F. (1987). Cell 50:17.
Jacquier, A., Rodriguez, J. R., and Rosbash, M. (1985). Cell 43:423.
Jacquier, A., and Rosbash, M. (1986). Science 234:1099.
Jarrell, K. A., Dietrich, R. C., and Perlman, P. S. (1988a). Molec. Cell. Biol. 8:2361.
Jarrell, K. A., Peebles, C. L., Dietrich, R. C., Romiti, S. L., and Perlman, P. S. (1988b). J. Biol. Chem. 263:3432.

Johnson, P. F., and Abelson, J. (1983). Nature 302:681.
Johnson, J., Ogden, R., Johnson, P., Abelson, J., Dembeck, P., and Itakura, K. (1981). Proc. Natl. Acad. Sci. USA 77:2564.
Kaine, B. P. (1987). J. Molec. Evol. 25:248.
Kaine, B. P., Gupta, R., and Woese, C. R. (1983). Proc. Natl. Acad. Sci. USA 80:3309.
Kang, H. S., Ogden, R. C., and Abelson, J. (1980). In Mobilization and Reassembly of Genetic Information. Scott, W. A., Werner, R., Joseph, D. R., and Schultz, J., eds. Academic Press, New York, p. 317.
Kao, T., Moon, E., and Wu, R. (1984). Nucleic Acids Res. 12:7305.
Karch, et al. (1985). Cell 43:81.
Kaufer, N., Simanis, V., and Nurse, P. (1985). Nature 318:75.
Kay, P. S., and Inoue, T. (1987). Nature 327:343.
Keller, E. B., and Noon, W. A. (1984). Proc. Natl. Acad. Sci. USA 81:7417.
Keller, M., and Michel, F. (1985). FEBS Letters 179:69.
Kim, S. H., and Cech, T. R. (1987). Proc. Natl. Acad. Sci, USA 84:8788.
Kjems, J., and Garrett, R. A. (1985). Nature 318:675.
Kjems, J., and Garrett, R. A. (1988). Cell 54:693.
Koch, W., Edwards, K., and Kossel, H. (1981). Cell 25:203.
Koller, B., Fromm, H., Galun, E., and Edelman, M. (1987). Cell 48:111.
Konarska, M. M., Grabowski, P. J., Padgett, R. A., and Sharp, P. A. (1985). Nature 313:552.
Konarska, M. M., Padgett, R. A., and Sharp, P. A. (1985). Cell 42:165.
Konarska, M. M., and Sharp, P. A. (1988). Cell, 46:845.
Kotylak, Z., Lazowska, J., Hawthorne, D. C., and Slonimski, P. P. (1985). In Achievements and Perspectives of Mitochondrial Research. Vol. II: Biogenesis, Quagliariello, E., Slater, E. C., Palmieri, F., Saccone, C., and Kroon, A. M. eds. Elsevier Science Publishers, Amsterdam, p. 1.
Krainer, A. R., and Maniatis, T. (1988). In Frontier in Transcription and Splicing, Hames, B. D., and Glover, D. M., eds. IRL Press, Oxford, p. 131.
Krainer, A. R., Maniatis, T., Ruskin, B., and Green, M. R. (1984). Cell 36:993.
Krause, M., and Hirsh, D. (1987). Cell 49:753.
Kreike, J., Schulze, M., Ahne, F., and Lang, B. F. (1987). EMBO J. 6:2123.
Kreike, J., Schulze, M., Pillar, T., Korte, A., and Rodel, G. (1986). Curr. Genet. 11:185.
Kruger, K., Grabowski, P. J., Zaug, A. J., Sands, J., Cottschling, D. E., and Cech, T. R. (1982). Cell, 31:147.
Kuch, U., Koppelhoff, B., and Esser, K. (1985) Curr. Genet. 10:59.
Kuiper, M. T. R., and Lambowitz, A. M. (1988). Cell 55:693.
Kurjan, J., Hall, B. D., Gillam, S., and Smith, M. (1980). Cell 20:701.
Labouesse, M., Dujardin, G., and Slonimski, P. P. (1985). Cell 41:133.
Labouesse, M., Herbert, C. J., Dujardin, G., and Slonimski, P. P. (1987). EMBO J. 6:713.
Labouesse, M., and Slonimski, P. (1983). EMBO J. 2:269.
Lambowitz, A. (1989). Cell 56:323.
Lamouroux, A., Pajot, P., Kochko, A., Halbreich, A. and Slonimski, P. (1980). In The Organization and Expression of the Mitochondrial Genome. Saccone, C. and Kroon, A. eds. North Holland Publishing Co., Amsterdam, p. 152.
Lang, B. F. (1984). EMBO J. 3:2129.
Langford, C. J., and Gallwitz, D. (1983). Cell 33:519.

Langford, C. J., Klinz, F. J., Donath, C., and Gallwitz, D. (1984). Cell 36:645.
Langford, C. J., Nellen, J., Niessing, J., and Gallwitz, D. (1983). Proc. Natl. Acad. Sci. USA 80:1496.
Lazowska, J., Jacq. C., and Slonimski, P. P. (1980). Cell 22:333.
Lee, M.-C., and Knapp, G. (1985). J. Biol. Chem. 260:3108.
Lemieux, C., and Lee, R. W. (1987). Proc. Natl. Acad. Sci. USA 84:4166.
Lin, R. J., Newman, A., Cheng, S. C., and Abelson, J. (1985). J. Biol. Chem. 260:14780.
Lustig, A. J., Lin, R. J., and Abelson, J. (1986). Cell 47:953.
MacPherson, J. M., and Roy, K. L. (1986). Gene 42:101.
Macreadie, I. G., Scott, R. M., Zinn, A. R., and Butow, R. A. (1985). Cell 41:395.
Majumder, A., Akins, R. A., Wilkinson, J. G., Kelley, R. L., Snook, A. J., and Lambowitz, A. M. (1989). Molec. Cell. Biol., in press.
Mannella, C., Collins, R. A., Green, M. R., and Lambowitz, A. M. (1979). Proc. Natl. Acad. Sci. USA 76:2635.
Mannella, C. A., and Lambowitz, A. (1979). Genetics 93:645.
Mao, J., Schmidt, O., and Soll, D. (1980). Cell 21:509.
McGraw, P., and Tzagoloff, A. (1983). J. Biol. Chem. 258:9468.
McPherson, J. M., and Roy, K. L. (1986). Gene 42:101.
Mecklenburg, K. L. (1986). Ph.D. Thesis, The Ohio State University.
Michel, F., and Dujon, B. (1983). EMBO J. 2:33.
Michel, F., Jacquier, A., and Dujon, B. (1982). Biochimie 64:867.
Michel, F., and Lang, B. (1985). Nature 316:641.
Michel, F., Umesono, K., and Ozeki, H. (1989). Gene, in press.
Miller, T. J., Stephens, D. L., and Merty, J. E. (1982). Molec. Cell. Biol. 2:1581.
Monaco, A. P., Neve, R. L., Colletti-Feener, C., Bertelson, C. J. Kurnit, D. M., and Kunkel, L. M. (1986) Nature 323:646.
Mota, E. M., and Collins, R. A. (1988). Nature 332:654.
Mount, S. M. (1982). Nucleic Acids Res. 10:459.
Mount, S. M., Pettersson, I., Hinterberger, M., Karmas, A., and Steitz, J. A. (1983). Cell 33:509.
Muller, F., and Clarkson, S. G. (1980). Cell 19:345.
Muller, M., Schweyen, R. J., and Schmelzer, C. (1988). Nucleic Acids Res. 16:7383.
Murphy, W. J., Watkins, K. P., and Agabian, N. (1986). Cell 47:517.
Muscarella, D., and Vogt, V. (1989). Cell 56:443.
Myers, A. M., Pape, L. K., and Tzagoloff, A. (1985). EMBO J. 4:2087.
Nargang, F., Bell, J. B., Stohl, L. L., and Lambowitz, A. M. (1984). Cell 38:441.
Neuhaus, H., and Link, G. (1987). Curr. Genet. 11:251.
Nishijura, K., Kurjan, J., Hall, B. D., and De Robertis, E. M. (1982). EMBO J. 1:263.
Nobrega, F. G., and Tzagoloff, A. (1980). J. Biol. Chem. 255:9828.
Ogden, R. C., Beckmann, J. S., Abelson, J., Kang, H. S., Soll, D., and Schmidt, O. (1979). Cell 17:399.
Ogden, R. C., Lee, M.-C., and Knapp, G. (1984). Nucleic Acids Res. 12:9367.
Olson, M. V., Page, G. S., Sentenac, A., Piper, P. W., Worthington, M., Weiss, R. B., and Hall, B. D. (1981). Nature 291:464.
Osiewacz, H. D., and Esser, K. (1984). Curr. Genet. 8:299.
Otsuka, A., de Paolis, A., and Tocchini-Valentini, G. P. (1981). Molec. Cell. Biol. 1:269.
Padgett, R. A., Grabowski, P. J., Konarska, M. M., Seiler, S., and Sharp, P. A. (1986). Ann. Rev. Biochem. 55:1119.

Padgett, R. A., Konarska, M. M., Grabowski, P. J., Hardy, S. F., and Sharp, P. A. (1984). Science 225:898.
Pape, L. K., Koerner, T. J., and Tzagoloff, A. (1985). J. Biol. Chem. 260:15362.
Parker, R., Siliciano, P., and Guthrie, C. (1987). Cell 49:220.
Partono, S., and Lewin, A. S. (1988). Molec. Cell. Biol. 8:2562.
Pearson, D., Willis, I., Hottinger, H., Bell, J., Kumar, A., Leupold, U., and Soll, D. (1985). Molec. Cell. Biol. 5:808.
Pedersen-Lane, D. and Belfort, M. (1987). Science 237:182.
Peebles, C. L., Gegenheimer, P., and Abelson, J. (1983). Cell 32:525.
Peebles, C. L., Perlman, P. S., Mecklenburg, K. L., Petrillo, M. L., Tabor, J. H., Jarrell, K. A., and Cheng, H. L. (1986). Cell 44:213.
Peffley, D. M., and Sogin, M. (1981). Biochem. 20:4015.
Perea, J., and Jacq, C. (1985). EMBO J. 4:3281.
Perkins, K. K., Furneaux, H. M., and Hurwitz, J. (1986). Proc. Natl. Acad. Sci. USA 83:887.
Phizicky, E. M., Schwartz, R. C., and Abelson, J. (1986). J. Biol. Chem. 261:2978.
Pikielny, C. W., Rymond, B. T., and Rosbash, M. (1986). Nature 324:341.
Pikielny, C. W., Teem, J. L., and Rosbash, M. (1983). Cell 34:395.
Pillar, T., Lang, B. F., Steinberger, I., Vogt, B., and Kaudewitz, F. (1983). J. Biol. Chem. 258:7954.
Price, J. V., and Cech, T. R. (1985). Science 228:719.
Price, J. V., Engberg, J., and Cech, T. R. (1987). J. Mol. Biol. 196:49.
Price, J. V., Kieft, G. L., Kent, J. R., Sievers, E. L., and Cech, T. R. (1985). Nucleic Acids Res. 13:1871.
Quirk, S. M., Bell-Pedersen, D., and Belfort, M. (1989). Cell 56:455.
Quirk, S. M., Bell-Pedersen, D., Tomaschewski, J., Ruger, W., and Belfort, M. (1989). Nucleic Acids Res. 17:301.
Ralph, D. (1986). Ph.D. Thesis. The Ohio State University, Columbus, Ohio.
Raymond, G., and Johnson, J. D. (1983). Nucleic Acids Res. 11:5969.
Reed, R., Griffith, J., and Maniatis, T. (1988). Cell 53:949.
Reed, R., and Maniatis, T. (1988). Genes Dev. 2:1268.
Reyes, V. M., and Abelson, J. (1988). Cell 55:719.
Robinson, R. R., and Davidson, N. (1981). Cell 23:251.
Rodriguez, J. R., Pikielney, C. W., and Rosbash, M. (1985). Cell 39:603.
Ruskin, B., and Green, J. M. (1985). Nature 317:732.
Ruskin, B., Green, J. M., and Green, M. (1985). Cell 41:833.
Ruskin, B., Krainer, A. R., Maniatis, T., and Green, M. R. (1984). Cell 38:317.
Ruskin, B., Zamore, P. D., and Green, M. R. (1988). Cell 52:207.
Schmelzer, C., and Muller, M. (1987). Cell 51:753.
Schmelzer, C., and Schweyen, R. J. (1986). Cell 46:557.
Selker, E., and Yanofsky, C. (1980). Nucleic Acids Res. 8:1033.
Seraphin, B., Boulet, A., Simon, M., and Faye, G. (1987). Proc. Natl. Acad. Sci. USA 84:6810.
Seraphin, B., Simon, M., and Faye, G. (1988). EMBO J. 7:1455.
Shapiro M. B., and Senapathy P. (1987). Nucleic Acid Res. 15:7155.
Shub, D. A., Yu, M.-Q., Gott, J. M., Zech, A., and Wilson, L. D. (1987). Cold Spring Harbor Symp. Quant. Biol. 52:193.
Siliciano, P., and Guthrie, C. (1988). Genes Dev. 2:1258.
Sjoberg, B. M., Hahne, S., Mathews, C. Z., Mathews, C. K., Rand, K. N., and Gait, M. J. (1986). EMBO J. 5:2031.

Slonimski, P., Pajot, P., Jacq, C., Foucher, M., Perrodin, G., Kochko, A., and Lamouroux, A. (1978). In Biochemistry and Genetics of Yeast, Bacilla, M., Horecker, B., and Stoppiani, A., eds. Academic Press, Orlando, FL, p. 339.
Sogin, M. L., Ingold, A., Karlok, M., Nielsen, H., and Engberg, J. (1986). EMBO J. 5:3625.
Solnick, D. (1985). Cell 42:157.
Sprinzl, M., Hartmann, T., Meissner, F., Moll, J., and Vorderwulbecke, T. (1987). Nucleic Acids Res. Suppl. 15:r53.
Standring, D. N., Venegas, A., and Rutter, W. J. (1981). Proc. Natl. Acad. Sci. USA 78:5963.
Stange, N., and Beier, H. (1986). Nucleic Acids Res. 14:8691.
Steinmetz, A., Gubbins, E. J., and Bogorad, L. (1982). Nucleic Acids Res. 10:3027.
Stohl, L. L., Collins, R. A., Cole, M. D., and Lambowitz, A. M. (1982). Nucleic Acids Res. 10:1439.
Strathern, J., Klar, A., Hicks, J., Abraham, J., Ivy, J., Nasmyth, K., and McGill, C. (1982) Cell 31:183.
Strobel, M., and Abelson, J. (1986a). Molec. Cell. Biol. 6:2663.
Strobel, M., and Abelson, J. (1986b). Molec. Cell. Biol. 6:2674.
Stucka, R., and Feldman, H. (1988). Nucleic Acids Res. 16:3583.
Sugita, M., Shinozaki, K., and Sugiura, M. (1985). Proc. Natl. Acad. Sci. USA 82:3557.
Sullivan, F.X., and Cech, T. R. (1985). Cell 42:639.
Sumner-Smith, M., Hottinger, H., Willis, I., Koch, T. L., Arentzen, R., and Soll, D. (1984). Molec. Gen. Genet. 197:447
Suter, B., Altwegg, M., Choffat, Y., and Kubil, E. (1986). Arch. Biochem. Biophys. 247:233.
Sutton, R. E., and Boothroyd, J. C. (1986). Cell 47:527.
Swerdlow, H., and Guthrie, C. (1984). J. Biol. Chem. 259:5197.
Szekely, E., Belford, H. G., and Greer, C. L. (1988). J. Biol. Chem. 263:13839.
Tabak, H. F., van der Horst, G., Kamps, A. M. J. E., and Arnberg, A. C. (1987). Cell 48:101.
Tabak, H. F., van der Horst, G., Osinga, K. A., and Arnberg, A. C. (1984). Cell 39:623.
Tazi, J., Alibert, C., Temsamani, J., Reveilland, I., Cathala, G., Brunel, C., and Jeanteur, P. (1986). Cell 47:755.
Teem, J. L., Abovich, N., Kaufer, N. F., Schwindinger, W. F., Warner, J. R., et al. (1984). Nucleic Acids Res. 12:8295.
Thomas, J. D., Conrad, R. C., and Blumenthal, T. (1988). Cell 54:533.
Thompson, L. D., and Daniels, C. J. (1988). J. Biol. Chem. 263:17951.
Trinkl, H., and Wolf, K. (1986). Gene 45:289.
Tschundi, C., Richards, F. F. and Ullu, E., (1986) Nuclear Acids Res. 14:8893.
Valenzuela, P., Venegas, A., Weinberg, F., Bishop, R., and Rutter, W. J. (1978). Proc. Natl. Acad. Sci. USA 75:190.
van der Horst, G., and Tabak, H. F. (1985). Cell 40:759.
van der Horst, G., and Tabak, H. F. (1987). EMBO J. 6:2139.
van der Veen, R., Arnberg, A. C., van der Horst, G., Bonen, L., Tabak, H. F., and Grivell, L. A. (1986). Cell 44:225.
van der Veen, R., Arnberg, A. C., and Grivell, L. A. (1987). EMBO J. 6:1079.
van Doren, K., and Hirsh, D. (1988). Nature 335:556.
van Santen, V. L., and Spritz, R. A. (1987). Gene 56:253.

Van Tol, H., and Beier, H. (1988). Nucleic Acids Res. 26:1951
Van Tol, H., Stange, N., Gross, H. J., and Beier, H. (1987). EMBO J. 6:35.
Venegas, A., Quiroga, M., Zaldivar, J., Rutter, W. J., and Valenzuela, P. (1979). J. Biol. Chem. 254:12306.
Waldron, C., Wills, N., and Gesteland, R. F. (1985). J. Mol. Appl. Genet. 3:7.
Wallace, J. C., and Edmonds, M. (1983). Proc. Natl. Acad. Sci. USA 80:950.
Waring, R. B., and Davies, R. W. (1984). Gene 28:277.
Waring, R. B., Ray, J. A., Edwards, S. W., Scazzocchio, C., and Davies, R. W. (1985). Cell 40:371.
Waring, R. B., Scazzocchio, C., Brown, T. A., and Davies, R. W. (1983). J. Mol. Biol. 167:595.
Waring, R. B., Towner, P., Minter, S. J., and Davies, R. W. (1986). Nature 321:133.
Watts, F., Castle, C., and Beggs, J. D. (1983). EMBO J. 2:2085.
Weiler, G. (1987). Ph.D. Thesis, University of Munich, Munich FRG.
Weiss-Brummer, B., Hall, J., Schweyen, R. J., Rudel, G. P., and Kaudewitz, F. (1983). Cell 33:195.
Weiss-Brummer, B., Rodel, G., Schweyen, R. J., and Kaudewitz, F. (1982). Cell 29:527.
Wenzlau, J., Saldanha, R., Butow, R. A., and Perlman, P. S. (1989). Cell 56:421.
Wich, G., Leinfelder, W., and Bock, A. (1987). EMBO J. 6:523.
Wickens, M. P. and Gurdon, J. B. (1983). J. Molec. Biol. 163:1.
Wiebauer, K., Herrero, J.-J., and Filipowicz, W. (1988). Molec. Cell. Biol. 8:2042.
Wierenga, G., Hofer, E., and Weissmann, C. (1984). Cell 37:915.
Williamson, C. L., Tierney, W. M., Kerker, B. J., and Burke, J. M. (1987). J. Biol. Chem. 262:14672.
Willis, I., Hottinger, H., Pearson, D., Chisolm, V., Leupold, U., and Soll, D. (1984). EMBO J. 3:1573.
Winey, M., and Culbertson, M. R. (1988). Genetics 118:609.
Winey, M., Mendenhall, M. D., Cummins, C. M., Culbertson, M. R., and Knapp, G. (1986). J. Mol. Biol. 192:49.
Xiong, Y., and Eickbush, T. H. (1988). Mol. Biol. Evol. 5:675.
Zaita, N., Torazawa, K., Shinozaki, K., and Sugiura, M. (1987). FEBS Letters 210:153
Zaug, A. J., Been, M. D., and Cech, T. R. (1986). Nature 324:429.
Zaug, A. J., and Cech, T. R. (1986a). Science 231:470.
Zaug, A. J., and Cech, T. R. (1986b). Biochemistry 25:4478.
Zaug, A. J., Kent, J. R., and Cech, T. R. (1984) Science, 224:574.
Zeitlin, S., and Efstratiadis, A. (1984). Cell 39:589.
Zhu, H., Conrad-Webb, H., Liao, X. S., Perlman, P. S., and Butow, R. A. (1989). Molec. Cell. Biol. 9:1507.
Zhu, H., Macreadie, I. G., and Butow, R. A. (1987). Molec. Cell. Biol. 7:2530.
Zhuang, Y., and Weiner, A. M. (1986). Cell 46:827.
Zimmer, M., Welser, F., Oraler, G., and Wolf, K. (1987). Curr Genet. 12:329.
Zinn, A. R., and Butow, R. A. (1985). Cell 40:887.

7

Alternative mRNA Splicing in the Generation of Protein Diversity and the Control of Gene Expression

CHRISTOPHER W. J. SMITH, DAVID KNAACK,
AND BERNARDO NADAL-GINARD

The origin, role, and selective advantage of introns has remained something of a mystery since their discovery 10 years ago. Alternative splicing offers a rationale for their existence and might hold some interesting clues about gene evolution. Alternative pre-mRNA processing is a versatile mechanism by which multiple protein isoforms can be produced from a single gene locus. As such, it has been exploited from viruses to mammals as an economic means by which to generate protein diversity, made possible only by the existence of introns. The mechanism of alternative RNA splicing is, at present, not well characterized, though it does appear to share a number of common features with the process of constituitive RNA splicing. Indeed, it seems likely that both regulated alternative splicing and strict constitutive splicing may represent refinements of an ancestral stochastic form of alternative splicing. Cis sequence elements involved in alternative splicing appear to reside in both the introns and exons, as well as in the splice sites themselves. Moreover, in at least some forms of tissue-specific and developmentally regulated alternative splicing, there is also trans regulation contributed by the specific cell environment in which splicing takes place. The scope of alternative splicing in affecting protein function extends from the determination of cellular localization to the deletion of functional activity. Furthermore, alternative processing of the untranslated regions of mRNA gives the potential for regulation of gene expression via the modulation of RNA stability or translational efficiency. Thus, differential RNA processing provides a powerful posttranscriptional mechanism by which both qualitative and quantitative regulation of protein expression can be achieved.

In addition, the ability to splice nonconstitutively may have played an important evolutionary role by reducing the genetic load of intragenic duplications and thus facilitating the selective utilization of duplicated exon sequences. An understanding of the basis of alternative mRNA splicing may ultimately cast light onto a number of fundamental problems in developmental and evolutionary biology.

MECHANISMS OF RNA PROCESSING

The primary eukaryotic mRNA transcript undergoes a series of processing and transport steps before it is eventually translated. While transcription is still under way the nascent transcript is "capped" by the covalent addition of a 7-methyl guanosine triphosphate base (Salditt-Georgieff et al., 1980). Transcription terminates well beyond the site of the mature 3' end of the transcript. The 3' end processing system then cleaves the transcript shortly after the consensus AAUAAA signal sequence and adds a poly (A) tail, which is typically 200–300 A residues long (Birnstiel et al., 1985). The 3' end processing generally precedes the excision of intervening sequences and ligation of exons that together constitute the splicing reaction. This sequence of events is probably due to the fact that 3' end processing is an intrinsically more efficient reaction than splicing, rather than to a requirement for a polyadenylated splicing substrate (Zeevi et al., 1981). The two major steps in the processing of the pre-mRNA, 3' end formation and splicing, take place in the nucleus, involve several steps, and occur in large complexes containing both protein and RNA components (Fig. 7.1; for recent reviews see Platt, 1986. Maniatis and Reed, 1987; Sharp, 1987; Birnstiel et al., 1985). 3' end processing involves the formation of a complex with protein and RNA components (Mowry and Steitz, 1987) and is dependent upon the presence of the obligatory AAUAAA signal (Proudfoot and Brownlee, 1976) and some other downstream sequence elements (Birnstiel et al., 1985; Gil and Proudfoot, 1987). Cleavage of the transcript and subsequent addition of a poly(A) tail then take place in two successive reactions (Fig. 7.1a). Splicing requires conserved sequence elements at the 5' "donor" site, the 3' "acceptor" site, and the "branch point." RNA and protein factors binding to these elements form a splicing complex known as the spliceosome (Sharp, 1987; Maniatis and Reed, 1987). Splicing involves two successive transesterification reactions (Fig. 7.1b). First, the 2'OH group of the branch point adenosine makes a nucleophilic attack upon the phosphate group between the upstream exon and the intron. As a result, the pre-mRNA is cleaved at the 5' donor site and the 5' guanosine of the intron becomes covalently joined to the 2'OH of the branch point adenosine residue. In the second step, the 3' splice site is cleaved and the two exons are ligated, yielding the reaction products, the spliced RNA and a branched "lariat" form of the intron. The successive excision of introns does not follow any strict directional order although there is evidence that a given transcript does tend to excise introns in an order that appears to be kinetically determined, some introns being more rapidly removed

164 INTERVENING SEQUENCES IN EVOLUTION AND DEVELOPMENT

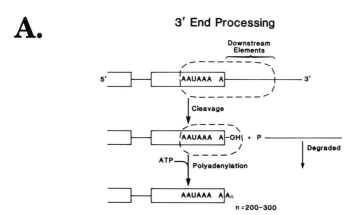

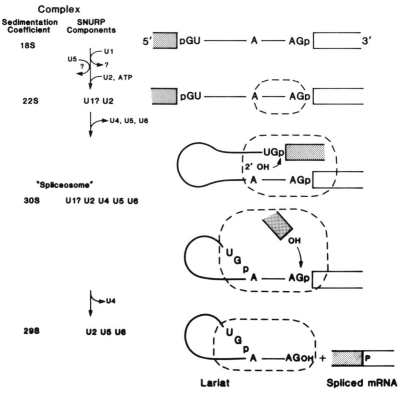

FIGURE 7.1. Mechanisms of pre-mRNA processing. (A) 3' end processing. The reaction occurs in two steps. A complex (denoted by the dashed lines) forms on the mRNA (exons are represented by blocks; introns and downstream sequences are represented by solid lines) dependent upon the presence of the AAUAAA consensus polyadenylation signal (Skolnick-David et al., 1987; Zarkower and Wickens, 1987). The pre-mRNA is then cleaved about 20 nt downstream of the AAUAAA

ALTERNATIVE mRNA SPLICING

than others (Lang and Spritz, 1987; Bovenberg et al., 1986). The spliced, polyadenylated, and capped mRNA can then be transported from the nucleus to the cytoplasm for translation. The two processing reactions are summarized schematically in Figure 7.1. Both of these processes can also be regulated in a differential fashion; alternative splicing and alternative poly(A) addition site selection are both used to generate different mRNAs from identical primary transcripts.

A fundamental problem confronting our understanding of the process of constitutive splicing is how the correct combinations of splice donor and acceptor sites are efficiently and accurately spliced together. Most mRNA transcripts contain, in addition to a larger number of bona fide donor and acceptor sites, many "cryptic" splice sites. These cryptic sites contain sequences that appear to conform to the consensus for donor and acceptor sites, yet they are never used, except in some cases when an adjacent splice site has been suppressed (Reed and Maniatis 1985). Thus, for constitutive splicing we need to know how, with so many available potential combinations of splice sites, only the correct pairs are ligated together in the mature transcript without skipping any one of them. Clues to this problem lie in the observations that adjacent exon sequences affect the efficiency with which splice sites are used (Reed and Maniatis, 1986), as does the degree to which they conform to the defined consensus sequences (Eperon et al., 1986: Zhuang et al., 1987). In addition, it seems that different pairs of splice sites may demonstrate a hierarchy of compatibilities with each other, which has evolved to ensure that only the correct pairs of splice sites are joined in the process of constitutive splicing. Constitutively spliced transcripts would have developed adjacent pairs of strong splice sites to ensure correct splicing at all times. In support of this notion is the finding that

---

signal. The efficiency of this reaction is dependent upon the presence of sequences downstream of the cleavage site. The addition of a 200–300-poly(A) tail then takes place, dependent upon the presence of the AAUAAA signal.

(B) The splicing reaction is shown in terms of the complexes formed with snRNPs (on the left) and in terms of the transesterification reactions (on the right). The approximate locations of the complexes on the substrate RNA are denoted by the dashed lines. In each case the upstream exon is shown as a filled block; the downstream exon, as an open block; and the intron, as a solid line. The GU of the 5' donor site, the A of the branch point, and the AG of the 3' acceptor site are all indicated. The initial steps involve complex formation. Although the U1 snRNP is known to have an early interaction with the 5' donor site, it is not clear whether it subsequently forms part of the stable "spliceosome" complex. Upon formation of the spliceosome (containing U2, U4, U5, and U6 snRNPs as well as the pre-mRNA) the first reaction involves the cleavage of the phosphodiester bond between the 5' exon and the intron and the formation of a 2'-5' phosphodiester between the branch point A and the donor site G. The second reaction involves cleavage of the bond between the 3' exon and the intron and the ligation of the two exons. The reaction products are the spliced exons and the branched "lariat" form of the intron. (Adapted from Konarska and Sharp, 1987; Sharp, 1987.)

PATTERNS OF ALTERNATIVE RNA SPLICING

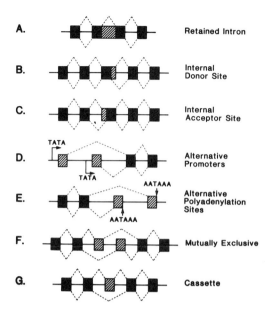

FIGURE 7.2. Patterns of alternative splicing. In each case, constitutively spliced exons are shown as solid blocks, and alternatively spliced exon sequences are shown as diagonally striped blocks. Introns are shown as solid lines, and splicing patterns are denoted by dashed lines. In the case of alternative promoters (D) and poly(A) sites (E), respectively, the TATA and AATAAA consensus sequences are shown.

single base substitutions within the constitutively spliced dihydrofolate reductase gene can lead to complex alternative splicing patterns (Mitchell et al., 1986).

The same basic problem of correct pairing of splice sites has to be addressed in the case of regulated alternative splicing. In this case, however, the situation is somewhat more complex, as we need to account for how the same splice sites can be accurately and efficiently spliced in different combinations in response to specific cellular environments and/or the structure of the primary transcript.

TYPES OF ALTERNATIVE PROCESSING: MECHANISTIC IMPLICATIONS

A number of different modes of alternative splicing can be distinguished according to the different combinations of splice sites that can be used (summarized in Fig. 7.2.). The simplest cases involve introns that are either excised or retained in the mRNA (Fig. 7.2A). Here the alternative is simple: whether or not to splice a given exon. Next is the use of alternative 5' donor (Fig.

7.2B) or 3' acceptor sites (Fig. 7.2C) on one side of an exon. Alternative utilization of these sites leads to the excision of introns of variable length, depending on the specific site used. In these first three examples, the distinction between exon and intron sequences is particularly diffuse, since a given sequence can be retained as an exon or excised as an intron sometimes concurrently in the same cell type. In these cases the possibility arises for alteration of the downstream reading frame, depending upon the pattern of splicing. The use of alternative donor or acceptor sites is particularly prevalent in some of the viral alternatively spliced genes, such as the adenovirus transcription units and the SV40 T antigen (Ziff, 1982). In other cases, processing involves the alternative use of discrete exons. Mutually exclusive 5' and 3' exons arising from differential usage of alternative promoters (Fig. 7.2D) or in conjunction with different exons containing poly(A) addition sites (Fig. 7.2E) have been documented in a number of different genes. In contrast, mutually exclusive splicing of internal exons (Fig 7.2F) has only been observed in α- and β-tropomyosins (Ruiz-Opazo and Nadal-Ginard, 1987; Wieczorek et al., 1988; Helfman et al., 1986) myosin light chain 1/3 (Periasamy et al., 1984; Strehler et al.,1985; Nabeshima et al., 1984), troponin-T (Medford et al., 1984), and M1 and M2 pyruvate kinase (Noguchi et al., 1986). Finally some "cassette" exons can be alternatively spliced independently of each other (Fig. 7.2G). In this case, the inclusion or excision of each exon does not appear to be dependent upon the splicing of any others. Such a mechanism has been shown to underlie the hypervariability of the N-terminal regions of fast muscle troponin-T (Breitbart et al., 1985).

The "Default" Splicing Pattern

In the cases of both splicing and 3' end formation, alternative processing presents an additional layer of mechanistic complexity to a basic system that is already highly complex and only partially understood. The further layer of sophistication in regulated alternative splicing may represent a refinement of constitutive processing or of a less precise stochastic ancestral splicing mechanism (see the following discussion). There is some evidence for the involvement of both subtle alterations in constitutive mechanisms and for the superimposition of specific mechanisms in alternative processing. It is beginning to appear as though a number of fundamental principles underlie the various modes of alternative splicing, although some of the forms are clearly more complex than others. One useful concept that has arisen from observations of the splicing behavior of native genes, transiently transfected gene constructs, and from transgenic mice, is the notion of a "default" splicing pattern. In a large number of cases it appears that alternatively spliced genes are spliced identically in a wide variety of cell and tissue types except for a small set of specific celltypes where an alternative splicing pattern is used. It would therefore appear that there is a default splicing pattern that will result from splicing by a constitutive splicing machinery and that is dictated in cis by a hierarchy of compatibilities between splice sites. This pattern of exon utilization would represent the basal, "unregulated" splicing of this particular transcript. Production of the alternative splicing pattern then requires the trans environment of a specific cell type, which presumably serves to modulate the hierarchy of compatibilities. This would, therefore, constitute the "regulated" or specialized pattern of splicing.

Alternative Processing of Transcripts with Different 5' Ends.

In principle, the regulated production of the different transcripts from a gene that has two differentially activated promoters, each with its associated 5' end exon, could be dictated entirely in cis. On the other hand, an additional contribution from the trans environment of specific cell types may be required. There are a number of examples in which alternative splicing occurs as a result of the differing primary structures of the pre-mRNAs, either because of the use of different promoters or because of cleavage and poly(A) addition at different exons. In these cases, since alternative splicing occurs upon transcripts that already have different 5' or 3' ends, there is no need to invoke any trans contribution to the regulation of alternative splicing. The alternative splicing patterns could be dictated entirely by the primary or secondary structures of the different transcripts. This appears to be the case with the myosin light chain 1/3 (MLC1/3) gene, which generates two different proteins by a mechanism of alternative splicing dictated by the differential activation of two different promoters located more than 10 kb apart (see Fig. 7.2d; Periasamy et al., 1984; Nabeshima et al., 1984; Strehler et al., 1985). In this case, activation of the 5' most promoter always leads to the production of MLC1 by the splicing of exon 1 to 4, bypassing exons 2 and 3, and subsequent constitutive splicing of exons 5–9. On the other hand, if the promoter immediately upstream of exon 2 is activated, this exon splices to exon 3, exon 4 is bypassed, and exons 5–9 are again constitutively spliced. The fact that constructs containing these exons are spliced identically in muscle and nonmuscle cells (M. Gallego, unpublished) confirms that no cell-specific trans regulation of alternative splicing occurs in this gene. Rather, the production of MLC1 or MLC3 is dictated by the differential activation of their specific promoters in front of exons 1 and 2, respectively. Exon 1, when present, is preferentially spliced to exon 4, and in its absence exon 2 is able to splice to exon 3. The differing primary structures of the two resultant transcripts are then able to specify, in cis, their specific splicing patterns.

Alternative 3' End Exons

The genes in which alternative processing involves alternative 3' end exons are among the best studied of alternatively processed genes. The use of alternative 3' ends involves two or more discrete poly (A) addition signals, in conjunction with alternative 3' end exons, for example, calcitonin–CGRP, Igμ_m and Igμ_s α- and β-tropomyosins (Amara et al., 1982; Alt et al., 1980; Early et al., 1980; Rogers et al., 1980; Ruiz-Opazo and Nadal-Ginard, 1986; Helfman et al., 1986). Thus, there is usually a mutually exclusive choice between use of a proximal or a distal exon each containing a poly(A) addition site. In principle, regulation of the production of one or another isoform in a tissue-specific or developmental fashion could be achieved at a number of points.

Transcription termination could selectively occur prior to the downstream poly(A) site, thus forcing the use of the internal site. Such a mechanism has been claimed for the Igμ gene in some cells (Kelley and Perry, 1986; Galli et al., 1987), although the assay used could not rule out transcription termination beyond the downstream poly(A) site with subsequent cleavage and poly(A) addition at the internal site and selective degradation of the 3' cleavage product (Galli et al., 1987). Most reports have claimed that transcription of the Igμ and the calcitonin–CGRP genes proceeds past the downstream poly(A) site even when the mature transcript utilizes the internal poly(A) site (Amara et al., 1984; Danner and Leder, 1985; Kelley and Perry, 1986; Reuther et al., 1986). If this is the case, alternative processing would always occur from the same primary transcript.

Alternative poly(A) addition sites appear to have different "strengths." Presumably, the distal site, the use of which does not necessarily preclude the subsequent use of the proximal site, would have to be stronger, or it would never be used. This has been shown to be the case in the Igμ gene (Galli et al., 1987) and is probably the case for the calcitonin–CGRP gene. The downstream μ_m poly(A) site appears to be used much more efficiently than the internal μ_s site. Nevertheless, the default splicing pattern in these two examples involves the use of the internal poly(A) site. The use of the internal poly(A) site could be up-regulated, either by increased levels of a constitutive polyadenylation factor(s), which is limiting in cells that use only the downstream site, or by the production of a factor specific for the proximal poly(A) site. If such regulation does occur, it is not a simple on–off switch, since both poly(A) sites in the Igμ gene can be used efficiently in all cells tested when they are not linked in cis, but are on separate constructs (Peterson and Perry, 1986; Galli et al., 1987).

The same end result could be obtained if instead of regulating the choice of the poly(A) addition site directly, it is the efficiency of the splice to the 3' distal exon that is specifically regulated; either "up" in cells that use the downstream site or "down" in cells that use the internal site. This type of regulation is almost certainly operative in the alternative processing of the calcitonin transcript. When gene constructs with non-tissue-specific promoters were expressed in transgenic mice, calcitonin, which results from the use of the internal poly(A) addition site, was found to be the "default" product, expressed even in tissues that do not normally express this gene. CGRP production was restricted to neuronal tissue (Crenshaw et al., 1987), suggesting that some specialized processing machinery is present in these cells. Furthermore, mutation of the internal calcitonin poly(A) addition site in transiently transfected constructs did not allow for the production of CGRP except in cells that normally express it, suggesting that the splice to the 3' exons coding for CGRP is specifically up-regulated in these cells, perhaps by the presence of a specific splicing factor (Leff et al., 1987). Similar results have been obtained with the Igμ gene.

Transfection experiments show that the internal μ_s poly(A) site seems to be the "default" choice and that mutation of the splice donor allows for the use of the μ_s poly(A) site in all cells. In contrast, mutation of the μ_s poly(A) site does not allow efficient splicing to give the μ_m product in inappropriate cell types (Tsuruchita et al., 1987). Again, this alteration in use of splice sites could be regulated either by changes in constitutive factors or by the production of specific splicing factors in early B-lymphocytes. Thus, in both of these examples it appears that it is the splice to the exon containing the downstream poly(A) site that is up-regulated in specific cell types and that therefore determines the structure of the mRNA.

No mechanistic studies have been carried out upon the alternative 3' end processing of the various tropomyosim genes (Helfman et al., 1986; Ruiz-Opazo and Nadal-Ginard, 1987). Though it is likely that the same general principles may be involved in these cases, the precise mechanism must be somewhat different. In contrast to the preceding examples in which use of the upstream poly(A) site is the default choice, in all the alternatively spliced tropomyosin genes it is the downstream poly(A) site that appears to be used as a "default" choice, whereas the internal poly(A) site is only used in striated muscle cells. Thus far there is no available information on the nature of the cis sequences that render the one poly(A) site more efficient than the other. However, it is likely that they involve sequences downstream from the cleavage site (Gil and Proudfoot, 1987).

Trans-*Acting Environments*

Regulated alternative splicing of transcripts with the same 5' and 3' ends requires the existence of cell-specific trans-acting environments. There are a number of instances in which identical primary transcripts are differentially spliced according to the cell type in which they are expressed. This is the case for the mutually exclusive and "combinatorial" exons of the rat fast muscle troponin-T gene (Medford et al., 1984; Breitbart et al., 1985), and the mutually exclusive exon pairs 2 and 3, as well as 7 and 8, of the rat α-tropomyosin gene (Wieczorek et al., 1988). In these instances, an identical primary transcript is alternatively spliced, depending upon the cell type or developmental stage where the transcript is expressed, demonstrating some trans contribution from the cell environment to the regulation of alternative splicing.

In the case of the rat α-TM gene (see Fig. 7.3) there are three regions of mutually exclusive splicing. One of each pair of mutually exclusive exons appears to represent a "default" choice for the splicing machinery. Whereas the other member of each pair requires a specific cell environment in order to be incorporated (Wieczorek et al., 1988). At the 5' end, exon 3 is used predominantly in all cells except for smooth muscle, where exon 2 is specifically incorporated. In the middle of the gene, exon 8 is the default choice used in all cells except in minor RNA species and in transformed cells, which use exon 7 instead. At the 3' end of the gene, mutually exclusive splicing involves the use of different poly(A) addition sites associated with exons 12 and 13; exon 13 is used in all cells except for striated muscles, where exons 11 and 12 are incor-

ALTERNATIVE mRNA SPLICING

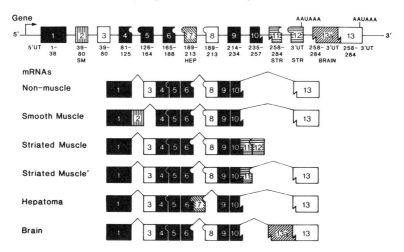

FIGURE 7.3. Rat α-tropomyosin gene and spliced mRNAs. The α-TM gene is split into at least 13 exons, the 3' most of which appears to have widely spaced alternative acceptor sites (D. Helfman, personal communication). The exons are shown as boxes with the following shading scheme: black—constitutive; white—default mutually exclusive; vertical stripes—smooth muscle specific; horizontal stripes—striated muscle specific; diagonal stripes upper left to lower right—hepatoma specific; diagonal stripes lower left to upper right—brain specific. The exons are numbered in numerical order 1–13 except that the sequences included by the use of the alternative brain-specific acceptor are labeled 13a. The default splicing mode is represented by the "non muscle" isoform, which uses exons 3, 8, and 13 from the three groups of mutually exclusive exons. Each of the other isoforms maintains two of these three alternative exons but varies one splicing choice. The exception is the minor striated muscle isoform that splices exon 11, a striated muscle 3' coding exon, to exon 13, rather than to exon 12, which contains the striated muscle-specific 3' UT sequence. (Adapted from Wieczorek et al., 1988.)

porated. In addition, two other mRNA species have been detected. A minor isoform in striated muscle appears to splice the striated muscle exon 11 to the "default" exon 13. In brain, exon 10 is spliced to an alternative upstream acceptor for exon 13 (D. Helfman, personal communication). The open reading frame (denoted exon 13a in Fig. 7.3) is terminated 26 codons downstream from the acceptor site and the remainder of exon 13a and 13 is treated as 3' untranslated sequence. These various spliced mRNAs are depicted in Figure 7.3. The "default" mRNA found in nonmuscle cells incorporates exons 3, 8, and 13. Interestingly, each cell type is able to activate the alternative incorporation of only one of the pair of exons without affecting the behavior of the other two pairs. The fact that each regulated member of the three mutually exclusive pairs is used in different cell types unambiguously demonstrates that distinct and specific trans environments are required for the incorporation of each one of them.

Expression of alternatively spliced constructs of Tn-T and α-TM in various cell types by transient transfections (Breitbart and Nadal-Ginard, 1987; A. Andreadis and authors' unpublished results) and of calcitonin CGRP in transgenic mice (Crenshaw et al., 1987) strongly supports the notion that a given transcript has a default mode of splicing that can be modified to give the alternative splicing pattern only in the appropriate specific cell type. In the case of the Tn-T gene, it has been demonstrated that the *trans*-acting factors in differentiated muscle cells are required to incorporate the combinatorial exons at the 5' end of the gene. In nonmuscle or undifferentiated muscle cells these exons are frequently bypassed and excluded from the mature mRNA. Moreover, the correct splicing activity is progressively induced during myogenesis (Breitbart and Nadal-Ginard, 1987). Although as yet there has been no clear-cut demonstration of the existence of specific alternative splicing factors, the available data are compelling. It should be pointed out, however, that the results obtained thus far could be equally well explained by modifications or differences in the levels of constitutive splicing factors. In either case, the obvious candidates for factors involved in alternative splicing would be the snRNPs (*s*mall *n*uclear *r*ibo*n*ucleo*p*rotein*s*). These ribonucleoprotein particles, which contain the U RNAs as well as a number of protein subunits (for review see Maniatis and Reed, 1987), are essential for pre-mRNA splicing (see Fig. 7.1). It is possible that heterogeneity of snRNPs, either within their RNA or protein components, could provide the basis for cell-specific trans regulation of alternative splicing.

Cis-*Acting Elements.*
During the past few years the notion has emerged that the sequences at or around the donor, acceptor, and lariat formation sites are necessary and sufficient to dictate the removal of the constitutively spliced introns (see Padgett et al., 1986; Sharp, 1987). The existence of alternative splicing has always been difficult to accommodate in this general hypothesis, especially since most alternatively spliced exons are flanked by donor, acceptor, and branch sites that do not deviate very significantly from the consensus sequences. Recently it has been demonstrated that exon sequences not immediately adjacent to the splice site can also affect its behavior (Reed and Maniatis, 1987).

So far, more progress has been made in the characterization of the cis elements, which may be involved in regulating alternative splicing, than of the putative trans factors. These cis elements may function either by determining the default splicing pattern or by acting as sites for interaction with the putative alternative splicing factors, which may alter the splicing pattern in specific cell types. Sequence elements in the introns and exons, and the splice donor and acceptor sites, have all been implicated. In the case of the Tn–T mutually exclusive exons, an intronic sequence 5' to the adult muscle-specific exon 16 but distant from both the donor and acceptor site has been implicated in preventing the incorporation of this exon in embryonic- or nonmuscle cells even when the alternative exon 17 has been deleted (A. Andreadis, unpublished). This intronic sequence may act as a "masking" element; in adult muscle cells that use exon 16, a specific factor may bind to this element, thus unmasking exon 16 and allowing its incorporation in place of exon 17.

The potential for complementary intronic sequence elements to sequester exons in hairpin and loop structures and thus cause exon skipping has been demonstrated (Solnick, 1985; Solnick and Lee, 1987). However, with such synthetically engineered hairpin structures the stem had to be in excess of 50 bp to cause skipping. This degree of complementarity is greater than that of any known naturally occurring hairpins, though it is possible that native hairpins could be stabilized by cellular factors. Complementary sequence elements have been found flanking the mutually exclusive exons of α-TM (Ruiz-Opazo and Nadal-Ginard, 1986) and Tn-T (Breitbart and Nadal-Ginard, 1986), which could form mutually exclusive hairpin structures. However, selective deletion of these elements in the α-TM gene had no effect upon the splicing pattern of transfected constructs, at least in cells that utilized the default splicing pattern (authors' unpublished results), suggesting that such secondary structural features do not underlie mutually exclusive splicing.

The involvement of an 81-bp exon sequence in the alternative splicing of the cassette EDIIIA exon of human fibronectin has recently been documented (Mardon et al., 1987). The exon is only incorporated when this specific sequence is present. Deletion of the sequence, or replacement by its complement, results in constitutive skipping of the exon. This sequence is therefore a prime candidate as a target sequence for a putative alternative splicing factor.

The involvement of the splice donor and acceptor sites in dictating alternative splicing patterns has long been suspected. The only fully conserved nucleotides in the splice sites and branch points of higher eukaryotes are the GU of the donor, the AG of the acceptor, and the A of the branch point. Differences occuring in the other positions of the consensus sequences may modulate the efficiency with which these sites are used. The complementarity between the donor consensus site and the 5' end of the U1 RNA immediately suggests that donor sites closer to the consensus may be "stronger" and used more efficiently than those that diverge significantly from the consensus. This suggestion has been confirmed by cis competition assays in which the ability of duplicated donor sites to compete with a reference site correlated with their degree of homology with the consensus (Eperon et al., 1986). Moreover, mutation of the SV40 large T donor site toward the consensus improves the efficiency of its use at the expense of the small t donor site (Zhuang et al., 1987). Differences in splice site efficiencies may also be affected by the neighboring sequences; replacement of exon sequences with unrelated "spacer" sequences has been shown to reduce the efficiency with which adjacent splice sites are used (Reed and Maniatis, 1986).

In the case of mutually exclusive exons the splicing behavior of the two exons must somehow be coupled so that the two of them cannot be incorporated simultaneously. With the mutually exclusive exons of MLC1/3, α-TM, and Tn-T, one member of each pair is incorporated as a default choice even when the two exons are flanked by a common donor and acceptor pair from a heterologous gene. Deletion of the default exon in MLC1/3 or α-TM (but not Tn-T, in which an intronic sequence element prevents incorporation of the other exon; see preceding discussion) then allows the other exon to be incorporated. It, therefore, appears that there is nothing wrong with the nondefault exon in these cells, but that there is a competition between the two mutually exclusive exons for the flanking donor and acceptor sites and that the "winner" of this competition is the default exon. The basis of the mutually exclusive nature of splicing in these genes may reside in the intron between the two exons. This intron is in no case excised as a unit even though it contains functional donor, acceptor, and branch point sites and is well above the minimal intron length. This could be due to the presence of other splice sites, in cis, with which the sites at each end of the intron are more compatible and in the absence of which the intron would be spliced out as a unit. Alternatively, the two splice sites may be absolutely incompatible. Experiments with constructs that contain the mutually exclusive exons 2 and 3 of α-TM, and the intron between, but no other splice sites, suggest that the two splice sites are incompatible even in the absence of any cis competition. The basis of the inability of the two exons to splice to each other is the unusual location of the exon 3 branch point, 177 nt upstream of the 3' splice site. This is only 42 nt downstream of the exon 2 donor site, which is below the minimal donor-branch point separation required for the formation of a productive spliceosome (Smith and Nadal-Ginard, 1989). This steric blocking mechanism is not general, however, since the branch points between other pairs of mutually exclusive exons, while often distant from the 3' splice site, are not always so close to the donor as to prevent splicing (Gallego and Nadal-Ginard, unpublished observations, D. Helfman, personal communication). With the mutually exclusive nature of splicing enforced by a nonsplicing intron, a hierarchy of splice site "strengths" can be invoked to explain the observed default splicing patterns. Production of the alternative nondefault splicing pattern would then involve the modulation of the splice site hierarchy and competition between the exons by other elements, cis in the case of MLC1/3 and trans in the case of α-TM and Tn-T. Obvious candidates for specific alternative splicing trans factors would be the snRNPs. For instance, in each of the mutually exclusive pairs of exons in α-TM and Tn-T the default exon has a donor site that is closer to the consensus than the alternative exon. The two donor sites may, therefore, specify the splicing default choice. Minor tissue-specific U1 RNAs that varied within the donor complement sequence could then represent alternative splicing factors, causing the specific incorporation of the nondefault exons. In this respect, however, although U1 variants have been documented, the reported heterogeneity does not lie within the sequence complementary to the donor site (Lund et al., 1985). It is therefore possible that putative tissue-specific differences between snRNP particles may reside in their protein components.

PROTEIN STRUCTURAL CONSTRAINTS UPON ALTERNATIVE SPLICING

Alternative RNA processing provides a means by which specific regions of the mature mRNA can be varied while the remainder of the message is conserved in all circumstances. If the variable segments are within the protein coding region, the structure and function of the protein product will be affected accordingly. If the inclusion or excision of the alternatively spliced exons results in the maintenance of the same translational reading frame, the protein products will be identical within a set of constant domains differing only within specific variable (additional or exchangeable) domains. This is the case with most known examples of mutually exclusive and cassette exons. On the other hand, there are a number of examples of retained introns or competing acceptor or donor sites (see Fig. 7.2) in which alternative splicing results in the use of different downstream reading frames. In these cases protein sequences to the C-terminal side of the alternative splice are unrelated. The degree of similarity between the protein products therefore depends upon the location of the alternative splice site within the message. For example, the use of alternative acceptors within the first intron of the *Drosophila tra* gene produces either a functional mRNA for the *tra* gene product or a message with no extended open reading frame (Boggs et al., 1987). In contrast, retention of the final intron of the PDGF A-chain produces a protein with only 15 C-terminal amino acids more than the spliced form (Tong et al., 1987; Collins et al., 1987).

Although the number of genes that are known to undergo alternative RNA processing has expanded dramatically within recent years (see, e.g., Breitbart et al., 1987; Andreadis et al., 1987), there are relatively few examples where enough is known about the the protein product to be able to correlate the functional consequences of alternative splicing with specific changes in molecular structure and interactions. Some principles, however, are beginning to emerge. Cassette exons and mutually exclusive exons, including those coding exons that arise through the use of different promoters or poly(A) addition sites, usually code for protein segments that are involved in interactions either with other proteins or with nonprotein ligands. Alternative splicing can therefore be used to modulate or abolish these interactions. Since these segments are involved with such interactions, it would seem that alternatively spliced exons are more likely to code for surface loops than for regions that form the folded core structure of globular protein domains. This would appear to hold especially in the case of cassette exons; the surface regions would be expected to have far more tolerance for the inclusion or exclusion of peptide segments compared with the tightly packed hydrophobic core. In the case of mutually exclusive exons this additional constraint may not operate, since one of the alternative exons is always included. Also, a large number of the known examples of alternative splicing occur within proteins that have a predominantly fibrous structure, such as tropomyosin (Crick, 1953), troponin-T (Pearlstone et al., 1977), and fibronectin (Hynes, 1985). These proteins do not have the same constraints upon folding as globular proteins and perhaps are more easily able to accommodate alternatively spliced segments. At present there is very little structural evidence to substantiate these arguments, although it has been noted that intron–exon boundaries do tend to map at the surface rather than in the interior of globular proteins (Craik et al., 1982). In the case of the myelin basic protein, secondary structural predictions place one of the cassette-encoded segments (exon 2) into a surface loop; the other (exon 6), however, codes for a connecting α-helix and a strand of the predicted central β-sheet structure (de Ferra et al., 1985). It is clear that more detailed knowledge of the individual systems in which alternative splicing is known to operate will be required before the specific protein structural constraints upon alternative splicing can be elucidated.

FUNCTIONAL EFFECTS OF ALTERNATIVE SPLICING

There are now many known cases in which specific patterns of alternative RNA processing can be correlated with biological function. In a small subset of these cases a further correlation can be made with the specific molecular interactions affected. These examples can be usefully grouped according to the effect of alternative splicing upon the protein function:

1. Cellular or subcellular localization.
2. Deletion of function.
3. Modulation of function.
4. Products with different functions.

These groupings are not mutually exclusive and some genes could be placed under more than one heading. This classification does, however, help to bring into perspective the diverse means by which alternative splicing can be used to affect protein function.

Cellular or Subcellular Localization

In certain cases regulated alternative RNA processing can be used to specify the localization or subcellular routing of the protein product. The best-characterized examples are those in which a protein can be either membrane bound or secreted. If a hydrophobic membrane-binding region is present, the protein will be found on the cell surface; in the absence of this domain the protein is secreted. In this type of regulation, the change is usually in the C-terminal of the protein and is therefore often achieved through the use of alternative 3' exons and poly(A) addition sites. This is the case in the immunoglobulin μ heavy-chain gene. In early B-lymphocytes the Igμ is principally present as a membrane-bound form. Upon B-cell maturation, following antigen stimulation, there is a progressive loss of the membrane-bound Igμ_m with a concomitant increase in the production of the secreted Igμ_s pentamer (Alt et al., 1980; Early et al., 1980; Rogers et al., 1980). This switch is achieved by the alternative use of 3' end exons in conjunction with separate poly(A) addition sites (see earlier discussion). Membrane-bound Igμ_m is produced from the downstream poly(A) site, whereas use of the internal poly(A) site leads to production of Igμ_s.

The gene for the "decay accelerating factor" (DAF) appears to give rise to membrane-bound and secreted forms by a different mechanism (Caras et al., 1987). DAF is a glycoprotein that is anchored to cell membranes by a linkage to phosphatidylinositol and that protects host cells from the complement cascade system by binding activated complement fragments C3b and C4b. A soluble form of DAF is also found in many body fluids, although the function of this form is presently unknown. In this case the two products are produced from transcripts ending with the same 3' untranslated sequences and poly(A) addition sites, differing only by an internal 118 nucleotides according to whether the final intron has been excised or retained. The shorter, spliced, form has a short hydrophobic C-terminal segment that makes the linkage to the phosphatidylinositol in the membrane-bound form. In the longer mRNA the open reading frame continues through the intron and into the final exon, introducing a frameshift that results in a longer, hydrophilic, C-terminus. This longer mRNA is believed to give rise to a secreted form of DAF (Caras et al., 1987).

The gene for the platelet-derived growth factor (PDGF) A-chain also has a final intron that can either be excised or retained (Tong et al., 1987; Collins et al., 1987). Splicing out of the intron in normal endothelial cells produces a termination codon after a further three amino acids. Retention of the intron in transformed cells produces a sequence with 15, predominantly acidic, additional amino acids before a termination codon is reached within the intron. This longer form is able to dimerize, be secreted, and has mitogenic activity. The spliced RNA does not produce any secreted activity. Presumably, the shorter A-chain is defective in either self-assembly or secretion and requires co-expression of the PDGF B-chain to become biologically active.

The actin-filament-severing protein gelsolin exists as both a plasma and a cytoplasmic protein. The two mature proteins are identical except for a 25-aa N-terminal extension on the plasma isoform. They are expressed from the same gene by the differential use of promoters and alternative splicing of the 5' exons coding for the additional 25-aa of the plasma protein and a 27-aa signal peptide necessary for its export that is cleaved from the mature plasma protein (Kwiatkowski et al., 1986). No functional differences between the intra- and extracellular forms of gelsolin have been detected. In this case, alternative splicing in conjunction with differential promoter usage determines the compartmentation of the protein.

Some correlation between location and alternatively spliced domains has been noted in various fibronectins. Incorporation of the ED type III cassette in fibroblasts produces cell fibronectin, whereas its removal in liver gives rise to plasma fibronectin (Kornblihtt et al., 1984, 1985). As yet, however, this particular domain has no known cell-binding or extracellular matrix-binding activities.

In addition to the preceding examples it is quite possible that alternative splicing will be found to regulate the distributions of proteins between various subcellular compartments. The targeting of proteins to specific compartments has been shown to involve the use of leader sequences that are removed from the mature protein during or after transportation to the appropriate location. Thus, an N-terminal amphipathic helix, with basic residues on the polar side, has been found to specify the transport of nuclear-encoded proteins into the mitochondria (Roise et al., 1986; Heijne, 1986). It is possible that the distribution of proteins that are common to more than one cell compartment could be regulated by alternative splicing of such targeting sequences, in a manner similar to the regulated production of cytoplasmic and secreted plasma gelsolin.

Deletion of Function

Alternative splicing can in some cases be used to remove one or more functions from a protein product. In the case of multifunctional proteins such selective removal of individual functions can be rationalized. However, there are some examples of alternative splicing in which one of the products appears to have no function. In true examples of this class, therefore, the function of the alternative splicing is to act as a simple on–off switch, in much the same way as transcription is used to control gene expression. Care has to be taken, however, in consigning genes to this class; lack of a known function is as likely to be indicative of ignorance on the part of the consigner as of redundancy of the protein product.

One gene that does seem to be a genuine member of this group is the *Drosophila* transformer *(tra)* gene, which is involved in female somatic sexual differentiation. This gene gives rise to two distinct mRNAs; the larger 1.1-kb transcript is sex nonspecific, whereas the smaller 0.9-kb transcript is found in females only (Boggs et al., 1987). The difference between the two mRNAs arises from the use of different acceptor sites within the first intron. The intron in the nonspecific mRNA is 73 nucleotides long and gives rise to a transcript with no extended open reading frame; the longest open reading frame begins more than two thirds of the way through the RNA and is only 48 amino acids long. The female-specific transcript, in contrast, uses an acceptor site that results in the excision of a 248-nucleotide intron, giving a single, long, open reading frame that codes for a 22-kDa protein that presumably possesses *tra* activity.

Some proteins are known to have a number of functions, one of which can be removed in certain forms of the protein by alternative splicing. The adenovirus E1a protein is a good example in this category. E1a has a number of activities, including transcriptional repression, which correlates with cell-transforming activity, as well as distinct transcriptional activation and induction of cellular DNA synthesis. E1a isoforms encoded by the 13S RNA have all these functions, whereas the isoform encoded by the 12S RNA, which utilizes an alternative 5' proximal donor site, lacks transcriptional activation but retains the other functions. This activity resides in a discrete 49aa peptide encoded by the sequence between the two donor sites and incorporated only into mature 13S RNA (Lillie et al., 1986, 1987).

The *Drosophila* P-element transposase provides an example where alternative splicing only gives rise to one known functional product; the other splicing mode does produce a protein but with no known function (Laski et al., 1987; Rio et al., 1987). The transposase gene, like DAF (see earlier), has a final intron that can be retained or spliced out. The intron is spliced out in germline cells only, giving rise to an active 87-kDa transposase protein. In somatic cells the intron is retained and the resultant 66-kDa protein has no sequences corresponding to the final exon and no transposase activity. Both proteins do contain a putative DNA binding bihelical structural motif, suggesting that the 66-kDa protein may have a functional role; the alternative splicing does, however, delete the transposase activity from one form of the protein.

Modulation of Function

Alternative splicing as a means to modulate protein function may well turn out to encompass the majority of cases of alternatively spliced genes. The use of cassette or mutually exclusive exons often appears to be geared toward this end. Such alternatively spliced protein domains often seem to coincide with the sites of interprotein interactions. Thus, the alternatively spliced regions of both α- and β-tropomyosins are in the areas that are known to interact with troponin-I and troponin-T in striated muscle, and in the zones involved in overlap of adjacent tropomyosin dimers (Ruiz-Opazo et al., 1985; Helfman et al., 1986). In troponin-T the mutually exclusive exons are in the region that is thought to interact with tropomyosin and troponin-C (Medford et al., 1984; Pearlstone and Smillie, 1983), whereas the cassette exons are in an area that abuts the overlap zone of adjacent tropomyosin dimers (Flicker et al., 1982). In rat and human plasma fibronectins, various alternative splicing modes (alternative donors, acceptors, and retained introns) within the IIIcs region correlate with a melanoma-specific adhesion activity. Isoforms that include the tetrapeptide Arg–Glu–Asp–Val, coded by the 3'-most part of the intron, have melanoma binding activity (Humphries et al., 1986; Kornblihtt et al. 1985; Rogers et al., 1987; Schwarzbauer et al., 1987). As yet, none of the other alternatively spliced regions of fibronectin have been shown to modulate any specific binding activity. The family of neural cell adhesion molecules (NCAMs) has been shown to undergo alternative splicing to give isoforms with identical extracellular domains but different membrane binding and cytoplasmic domains (Cunningham et al., 1987). The extracellular domains contain the important binding activities and have some homology with immunoglobulin heavy chains. Two of the three major types have a membrane-spanning segment and one of these two also has a cytoplasmic domain. The third and smallest isoform does not have a membrane-spanning segment, but it does have an anchoring domain through which it is linked to phosphatidylinositol. In addition, a muscle cell NCAM-specific extracellular domain has recently been identified (Dickson et al., 1987). The functional significance of these alternatively spliced domains remains to be determined, although it is clear that the forms with cytoplasmic domains have the potential for regulating cytoplasmic events according to the adhesion state of the cell, or conversely, for cytoplasmic regulation of extracellular adhesion.

One example of alternative splicing where correlations can be made with modulation of function and with interprotein interactions is the case of the "combinatorial" exons of rat fast muscle troponin-T. These five cassette mini-exons appear to be incorporated into mRNA largely independently of each other, giving the potential to generate mRNAs with up to 32 different combinations (Breitbart et al., 1985). In fact, some of these species are far more abundant than others and the actual number of isoforms that are produced in significant amounts in muscles may be somewhat less than this maximum potential; in a wide range of adult muscles the number of troponin-T proteins present in significant amounts appears to be between 5 and 10 (Briggs et al., 1987). Nevertheless, the splicing pattern of these exons does appear to be functionally significant. The N-terminal region of the protein for which they code interacts with the region in which tropomyosin dimers overlap to form a continuous strand along the muscle thin filaments (Flicker et al., 1982). The switching on of fast muscle contraction by Ca^{2+} binding to troponin-C and subsequent movement of the tropomyosin on the thin filament is a highly cooperative process, with the whole thin filament being switched on in a concerted manner as a single unit (Brandt et al., 1984). The degree of cooperativity (as expressed in the Hill coefficient n_H) and the Ca^{2+} concentration required for half maximal tension development in skinned rabbit muscles can be correlated with the particular combination of tropomyosin dimers (α–β or α–α) and the three major troponin-T isoforms present (TnT1f, TnT2f, and TnT3f). In muscles that have the same tropomyosin content but differ only in the troponin-T isoforms, there is a direct correlation between the troponin-T isoforms and these parameters (Schachat et al., 1987). The variation between these troponin-T species is in the N-terminal region coded by the "combinatorial" exons, and the specific difference between TnT1f and TnT2f is in the inclusion or exclusion of exon 4 (Briggs et al., 1984, 1987). Thus, the differences in the splicing pattern in this region of the TnT gene have specific effects upon the interaction of TnT with tropomyosin, which in turn gives rise to muscles with distinct mechanical parameters in their activation by Ca^{2+}.

One of the only examples in which changes in enzymatic activity can be correlated with alternative splicing is provided by the M isoforms of the glycolytic enzyme pyruvate kinase (PK). PK has four isoforms—M1, M2, L, and R—each of which forms active homotetramers. The M1 and M2 isozymes of PK are derived from the same gene and differ only within an internal 45-aa segment which is coded for by a pair of mutually exclusive exons (Noguchi et al., 1987). The M1 isozyme differs from the other forms of PK in that the native homotetramer shows Michaelis–Menten kinetics with respect to the substrate phosphoenolpyruvate, whereas the other isoforms are all allosterically regulated by phosphoenolpyruvate, fructose-1,6-bisphosphate, ATP, alanine, and pH and have sigmoidal kinetics (Imamura and Tanaka, 1982). In general, the M2, L, and R isoforms are found in tissues where their regulation can be employed to prevent futile cycling of the common intermediates of glycolysis and gluconeogenesis. In contrast, the M1 isoform occurs in muscle where glycolysis predominates. The sequence of the alternatively spliced region of M1 is quite distinct from that of the other three isoforms, which have a high degree of homology in this region. The crystal structure of feline M1 PK has been determined by X-ray diffraction (Stuart et al., 1979) and has been correlated with the amino acid sequence (Muirhead et al., 1986). The alternatively spliced exons M1 and M2 encompass the regions involved in intersubunit contact, suggesting that the nature of the subunit interactions, determined by alternative

ALTERNATIVE mRNA SPLICING

splicing, dictates the kinetic and regulatory properties of the individual isozymes (Noguchi et al., 1987).

Products with Different Functions

There is a limited number of cases in which alternative RNA processing can give rise to mature protein (or peptide) products with totally different functions. For this to be the case, the alternatively spliced region must code for the major active site of the protein product. The calcitonin gene provides a good example of this functional class of alternative splicing, producing both calcitonin and "calcitonin gene-related peptide" (CGRP) by alternative splicing and poly(A) addition site usage (Amara et al., 1982). The first three exons code for the 5' untranslated and common coding regions and are constitutively spliced. Exon 4 contains the calcitonin coding sequences and a poly(A) addition site; this exon is used in thyroid C cells. In neural tissue exon 3 is spliced instead to exons 5 and 6, which contain the CGRP coding and 3' untranslated sequences and poly(A) addition site. The primary protein products therefore contain an identical N-terminal end with divergent C-termini. However, the N-terminus is removed by proteolytic processing and the mature peptides are encoded entirely by the variable sequences. Calcitonin functions as a circulating Ca^{2+} homeostatic hormone, whereas CGRP may have both neuromodulatory and trophic activities (Rosenfeld et al., 1984; Fontaine et al., 1987). The calcitonin and CGRP coding sequences do appear to have arisen from an initial duplication event (Rosenfeld et al., 1984). Nevertheless, the present calcitonin–CGRP gene is a locus that is able to generate entirely different biological activities by a process of tissue-specific mRNA processing.

The number of genes in which alternative splicing is known to generate mature products with modified or different activities is at present somewhat limited, but it is rapidly growing. Given the rate at which new genes are being found to belong to this class it is likely that whole new families of genes will be shown to employ this device to generate diverse products. The ability of alternative splicing to maintain the structure of large parts of a protein while making modular substitutions of specific domains that may be involved in intermolecular interactions could, for example, be used in the generation of transcriptional activation factors or hormone receptors. Such proteins have interactions with both modifying or target molecules, as well as with other proteins that they may activate. Alternative splicing would provide a versatile mechanism by which different regulatory sites (sequence-specific DNA binding, hormone binding) could be combined with different activating or repressing functions. For instance, the mechanism involved in activation of transcription is likely to be the same for a large number of such molecules whereas the interaction with a hormone or specific DNA sequence is likely to vary. Thus, it would appear that alternative splicing would be well suited to generate families of these important regulatory proteins by combining constant domains that contain an activating function with variable domains that would specify the different upstream DNA-binding or hormone-binding sites. Such behavior has already been demonstrated in the case of alternatively spliced thyroid hormone receptors (Izumo and Mahdavi, 1988) and has been postulated for the homeo box containing genes of the bithorax complex of *Drosophila* (Pfeifer et al., 1987).

ALTERNATIVE PROCESSING OF UNTRANSLATED REGIONS OF RNA: POSSIBLE REGULATORY ROLES

Following splicing, the mature mRNA transcript must be transported from the nucleus, targeted to either free- or membrane-bound polysomes, translated and finally degraded. Each of these steps has the potential to be regulated and thus to provide another means by which to regulate gene expression. The 5' and 3' untranslated (UT) sequences of mRNA have been implicated in regulating translational efficiency and mRNA stability, respectively. There are a large number of genes in which alternative processing of pre-mRNA gives rise to mRNAs with different 5' and/or 3' UT sequences, sometimes but not necessarily, in conjunction with alternative 5' or 3' coding regions. In these cases there is a clear potential to regulate levels or kinetics of gene expression. Such regulation could be used to modulate the expression of a single isoform if only the UT regions are varied or to achieve appropriate levels of expression of distinct isoforms if coding regions are alternatively spliced in conjunction with the UT sequences.

Alternative Splicing of 5' Untranslated Regions

A number of genes are now known that show heterogeneity in their 5' UT sequences. This can be a result either of alternative promoter utilization, or of alternative splicing of UT exons which are transcribed from the same promoter. In either case the potential arises to regulate the translational efficiency of the transcript. 5' UT sequences are thought to regulate translational efficiency by the sequence context that they provide for the initiator AUG codon, through secondary structural elements and through the presence of upstream AUG codons and open reading frames (Kozak, 1986). In yeast, the 5' UT regions of the mRNAs for amino acid biosynthetic enzymes contain multiple AUG codons and act as sites for feedback inhibition of translation (Fin, 1986; Mueller and Hinnebusch, 1986; Tzamarias et al., 1986). The genes for mammalian HMG CoA reductase (Reynolds et al., 1985), HMG CoA synthase (Gil et al., 1987), α-amylase (Young et al., 1981), argninosuccinate synthetase (Freytag et al.a, 1984) insulin-like growth factor II (Irminger et al., 1987), and the β-subunit of thyrotropin (Wolf et al., 1987) all show heterogeneity within the 5' UT sequences of mRNAs that encode identical proteins. HMG CoA reductase has a highly variable 5' UT region ranging from 68 to 670 nucleotides, with variable numbers (0–8) of upstream AUG codons. This heterogeneity results from multiple initiation of transcription sites and from multiple 5' donor sites for the intron within the 5' UT sequence (Reynolds et al., 1985). The mRNA for HMG CoA synthase, on the other hand, has heterogeneity within the 5' UT sequences because of the presence of a cassette exon within the 5' UT region. It has been speculated that the different 5' UT regions could be used in the modulation of feedback suppression of translation or in the selective cytoplasmic or mitochondrial targeting of the enzyme (Gil et al., 1987). For genes such as MLC1/3, which display developmentally regulated use of two promoters to produce isoforms with distinct 5' ends, appropriate levels of expression of the alternative isoforms might be achieved, in part, by different translational efficiencies of the two mRNAs.

Alternative Splicing of 3' UT Regions

The 3' untranslated regions of mRNAs are known to be involved in specifying mRNA stability (Brawerman, 1987; Raghow, 1987), presumably in part because they provide the point of entry for 3' exonucleases. A repeated AUUUA motif has been found in the 3' UT region of a number of rapidly turning over mRNAs. Insertion of this element into the 3' UT region of rabbit β-globin was shown to have a dramatically destabilizing effect (Shaw and Kamen, 1986). The colony-stimulating factor-1 (CSF-1) gene produces two transcripts that are identical in the coding region but that differ only in the 3' UT region; interestingly, one of these 3' UT regions contains the destabilizing AUUUA repeating element, suggesting that CSF1 expression can be regulated at the level of mRNA turnover according to which 3' UT sequence is utilized (Ladner et al., 1987). Variation in transcript stability in response to physiological conditions can also effectively regulate protein expression and can be modulated by elements in the 3' UT region. For instance, a repetitive "ID" sequence has been found in the 3' UT region of a number of transcripts and has been shown to increase the stability of these transcripts in response to growth and viral transformation (Glaichenhaus and Cuzin, 1987). Moreover, this sequence element has been found to be present in an alternatively spliced 3' UT region of the L isoform of pyruvate kinase, raising the possibility for differential regulation of L-PK expression mediated by mRNA stability (Lone et al., 1986).

EVOLUTIONARY ASPECTS OF ALTERNATIVE SPLICING

Alternative Splicing and Protein Diversity

Alternative splicing has a large capacity for generation of protein diversity at low genetic cost. The three main mechanisms used to produce protein isoform diversity are gene duplication, gene rearrangement, and alternative splicing. Of the three, gene duplication is the least economic in terms of genetic material, since the coding of each isoform requires a repetition of the entire coding sequence. Still, the abundance of multigene families (Maeda and Smithies, 1986) demonstrates the evolutionary success of this process. The characterization of two differentially regulated actin genes with a single aspartate to glutamate substitution (Romans et al., 1985) suggests that the ability to place copies of the same gene under separate regulatory pathways may, at least in part, contribute to the evolutionary importance of this process.

A more efficient mechanism for the generation of multiple isoforms is gene rearrangement, as exhibited by the immunoglobulin and T-cell receptor genes. These genes, like alternatively spliced genes, contain intragenic duplications that correspond to the variable domains of the protein. Combinatorial rearrangement of the V, D, and J exons allows an almost unlimited number of antibody combinations to be produced (Golub, 1987). Given its dramatic power for the economic generation of isoform diversity, it is surprising that this mechanism seems to be restricted to the immunoglobulin and T-cell receptor family. One possible explanation for this restriction is that DNA rearrangement appears to be a unidirectional process. It irreversibly changes the genetic content of the cell (Sakano et al., 1979; Seidman et al., 1980; Cory and Adams; 1980, Sakano et al., 1981; Okazaki et al., 1987), and therefore it can only be used by cells in the process of terminal differentiation. Moreover, its uniqueness and complexity become biologically significant because of another property of the cells using this mechanism: clonal expansion. The DNA rearrangements leading to the production of unique immunoglobulin or T-cell receptor molecules would be of little value if these cells were not able to be selected and clonally expanded in the adult organism in order to produce a clonal population of identical cells, sufficient to play a role in the immune response of the organism.

Alternative splicing offers most of the advantages and few of the limitations of DNA rearrangement and duplication. Like DNA rearrangement, alternative splicing offers the possibility to generate multiple protein isoforms from a single gene. For a gene with a number (n) of alternatively spliced exons, there are 2^n possible isoforms, assuming a combinatorial mechanism of exon usage. Thus, 20 combinatorial exons have the potential to generate over a million different isoforms. This potential is further increased in some genes, such as the GTP binding protein Gsα, by the ability to generate further diversity at exon borders using duplicated acceptor or donor sites in a manner very similar to the diversity created at the join site in the immunoglobulin gene (Homcy et al., submitted). Furthermore, the amount of genetic material required to code for these isoforms is considerably less than if they were generated through gene duplication. Troponin T, for example, has 18 exons, five of which are spliced in a combinatorial and two of which are spliced in a mutually exclusive fashion (Medford et al., 1984; Breitbart et al., 1985). Therefore, this 20 kb gene has the capacity to code for 64 different isoforms. To encode the same informational content in 64 different transcriptional units would require more than 1,000 kb.

The potential contribution to phenotypic variability is increased many fold in the case of multiprotein complexes, especially when the alternatively spliced exons are located in different genes. This is exemplified in the case of the sarcomere. In vertebrates, this structure is produced by the assembly of seven major contractile proteins, each encoded by a multigene family with a minimum of four members per family (Buckingham and Minty, 1983). In combination, these genes have the theoretical capacity to generate $5^7 = 78,125$ different sarcomeric types, assuming an average of five members per multigene family, and that all isoforms could be used in a combinatorial fashion. This potential is increased many fold by alternative splicing. To date, more than 25 different exons or pairs of exons have been shown to be alternatively spliced in contractile protein genes (Andreadis et al., 1987; Breitbart et al., 1987), raising to many billions the number of different sarcomeres potentially produced by this limited set of genes. In fact, this maximum potential is not realized, and expression of many of the protein isoforms is in some way coordinately regulated. Nevertheless, even if only a subset of the genes and alternatively spliced patterns is expressed in a given cell type, the potential for the generation of different sarcomeric types is still very large.

In contrast to DNA rearrangement, alternative splicing does not change the genetic content of the cell, nor is it irreversible in practical terms. Splicing pathways need not be irreversibly discarded to adopt different ones. This feature makes it particularly advantageous for generating isoform diversity early in development and in long-lived cells. Isoforms generated by either alternative splicing or gene duplication allow for changes in phenotypic expression without permanent changes in the genome that might restrict the potential for gene expression in daughter cells. The potential for reversible regulation may be even more significant for long-lived terminally differentiated cells. Such cells may need to adapt protein isoform expression to widely different physiological and pathological conditions occurring over the course of their lifetime. This may be the case for neurons and muscle cells, which have a life span close to that of the organism.

Alternative Splicing in Protein Evolution

Alternative splicing might be important in protein evolution by facilitating efficient exploitation of intragenic DNA duplications and exon shuffling. DNA duplication with subsequent divergence has been a major source of genetic variability and has played an important role in evolution (Ohno, 1970). Duplication of nucleotide sequences represents a continuum ranging from small intragenic duplications to large-scale duplications involving complete genes, chromosomes, or the whole genome. Duplications of gene units or larger are likely to prove advantageous or neutral in the long run if they do not have deleterious gene dosage effects. On the other hand, intragenic duplications or translocations, as postulated for genes formed through exon shuffling (Darnell 1978, Gilbert 1978), would be expected to have neutral or deleterious effects. Assuming that such duplications or translocations occur within a constitutively spliced gene, their precise effects will depend on how much and what region of a particular gene is affected. Clearly, events limited to the intron sequences might have little or no effect, whereas duplications or translocations resulting in disrupted or rearranged exons or splice sites could have considerable consequences for gene expression. Even if translational frame shifts, stop codons, and splicing defects were avoided the changes in the primary protein sequence could easily disrupt protein conformation and function. This scenario is in contradiction with the increasing evidence that exon duplication and shuffling have constituted a primary mechanism for protein evolution. The apparent discrepancy between the genetic cost of most intragenic gene rearrangements and their evolutionary success suggests a role for alternative splicing. If newly duplicated or translocated exons were alternatively spliced, then the mutated gene could continue to produce the old gene product in addition to transcripts arising from new splicing patterns.

The new transcripts arising from alternative splicing of pre-mRNA containing a pair of duplicated exons could result in the evolution of new protein isoforms. New isoforms could arise either from novel transcripts containing additional exonic sequences or from the pair of identical transcripts containing a single copy each of the duplicated exon. The presence of both exons together on the mature transcript will be selected for or against, based on the expression of the protein product they encode. Should the presence of both exons within the same processed transcript not be immediately more advantageous than the original (coordinately expressed) transcript, the success of the two-exon transcripts would be largely determined by the fate of the single-exon transcripts. That is, random base substitution would occur with an equal rate in each of the two exon copies, since selection pressures would be equal. Eventually one transcript containing one of the duplicated exons would become more or less advantageous than the other. Since the expression of the functional protein product could still be ensured by the expression of the other transcript, the less advantageous copy would become shielded from selection, in the sense that it would not affect the gene's function (cf. Hunkapillar et al., 1982). Under this relaxed selection the less advantageous copy would undergo comparatively rapid sequence drift, and the other copy, along with the remainder of the protein-coding sequence, would remain under selection because of its expression in the other transcript. The rapid drift of the duplicated exon could result in its deterioration into an intronic sequence, or if the coordinate expression of the transcripts is maintained, at some point the exon may offer a selective advantage when expressed within the framework of the more highly conserved remainder of the protein. This could be either as one of a pair of mutually exclusive exons or in a transcript with both exons expressed. At such time, the isoform could be subject to renewed selection and its expression could become fixed.

Origin of Alternative Splicing

Alternative splicing might be as old as, or a predecessor of, constitutive splicing. Splicing is a very old posttranscriptional process, since it is present in archeobacteria (Kaine et al., 1983) and T-even phages (Chu et al., 1984, 1986). It might have evolved from autocatalytic RNA ligation, as demonstrated by the existence of self-spliced introns (Cech, 1986; Cech and Bass, 1986). The question arises as to whether regulated alternative splicing preceded or has evolved from constitutive splicing. The arguments presented in the previous section assume that an unregulated form of alternative splicing is also a very old process. Moreover, the demonstration that the distinction between constitutive and regulated alternative splicing is a relative one, not exclusively determined by the intrinsic properties of the exon and its splice sites (Breitbart and Nadal-Ginard, 1987), also argues that the two processes might be a part of a continuum and might have evolved simultaneously. If, as is widely believed, most present-day genes have been assembled through exon shuffling and duplication (Darnell, 1978; Gilbert, 1978), it is likely that the initial coding sequences carried out some primordial function. It is tempting to speculate that the new exon combinations were first alternatively spliced. The alternative sites might evolve into constitutive ones through mutation only if they are advantageous and thus become fixed in their splicing pattern. According to this scenario, stochastic alternative splicing would be the primitive splicing mechanism, and constitutive and regulated alternative splicing would represent its further refinement.

Unfortunately, the evolutionary evidence in support of this hypothesis is not available. However, from the data presently available it is clear that most alternative splicing occurs among exons that originated through duplications. Whether these duplications are recent or were present in the ancestral gene is open to question. Structural comparisons among myosin light-chain (Periasamy et al 1984), troponin-T (Breitbart and Nadal-Ginard 1986), and α-TM (Ruiz-Opazo and Nadal-Ginard 1987) genes suggests that most of these duplications are ancient. Perhaps the best-documented example in support of this hypothesis is offered by the α-TM gene, whose primary structure and splicing pattern is available from insects, birds, and mammals (Wieczorek et al. 1988). Of the 13 exons of this gene, only five are not alternatively spliced in one or another species. The splicing pattern of present-day TM genes could be the result of recent exon duplications. Alternatively, these duplications could have arisen early in evolution, with the particular organization and splicing pattern characteristic of each gene having evolved subsequently. Assuming equal rates of sequence drift, alternatively spliced exons, being the result of recent duplications, should exhibit a higher degree of intraspecies rather than interspecies conversation. In the second case, if all or most of the alternatively spliced exons were already present in the ancestral gene, the opposite result would be expected. The comparison of the different TM genes supports the latter conclusion and suggests that all the pairs of mutually exclusive exons found so far in TM genes were already present in the ancestral gene before the radiation of the insects more than 600 million years ago (Ruiz-Opazo and Nadal-Ginard, 1987). This hypothesis, if correct, implies that the ancestral TM gene present at the time of radiation of insects had a more complex structure than current TM alleles. Given such an organization, the pattern of alternative splicing must have been at least as complex as that of the present-day rat α-TM gene (Wieczorek et al., 1988). The possibility that the exons were constitutively spliced to produce a longer TM molecule seems remote, given the number, nature, and conservation of split codons at the end of alternatively spliced exons. Constitutive splicing of these exons would generate stop codons and frameshifts, resulting in either truncated TMs or proteins of unrelated primary and secondary structure. The data, therefore, suggest that, at least for the TM gene, new exons generated through duplication were first alternatively spliced.

The preceding hypothesis is based upon the assumption that the two members of a pair of duplicated exons are at all times under similar selective pressures. A significantly higher rate of sequence drift in one of a pair of more recently duplicated exons, during a period of relaxed selection, would also result in a greater interspecies exon sequence conservation (see earlier). This scenario, like the preceding hypothesis, requires the existence of alternative splicing at the time of the duplication event.

From these arguments we would like to suggest that alternative splicing either preceded or evolved concomitantly with constitutive splicing and that it has played an important role in gene evolution that might continue to the present day. Genes that need to function in different cell environments could have retained alternative splicing as a means of optimizing the product to the particular conditions of the cell. In contrast, genes that are expressed in a single cell type or whose products have already been optimized for a given cell environment or function would tend to contain constitutive exons and possibly reduced numbers of introns. Furthermore, for genes coding for proteins that precisely assemble into multiprotein complexes, such as the sarcomere or the cytoskeleton, the advantages of alternative splicing could have outweighed those of gene duplication with subsequent specialization. Alternative splicing facilitated the evolution of families of protein isoforms with constant and variable regions, domains critical for function being comprised of exons with fixed patterns of expression, whereas those important for the modulation of function or intermolecular interactions were alternatively spliced. This hypothesis is consistent with the low level of alternative splicing in lower organisms and its increased prevalence and complexity in higher metazoans.

NOTE ADDED IN PROOF

Alternative splicing is a rapidly expanding field of investigation. Since the time of writing this chapter, a number of developments have occurred. We refer interested readers to Smith et al. (1989) and references therein, for more recent developments, particularly with regard to the underlying mechanisms of alternative splicing, and its role in developmental regulation.

REFERENCES

Alt, F. W, Bothwell, A. L. M., Knapp, M., Siden, E., Mather, E., Koshland, M., and Baltimore, D. (1980). Synthesis of secreted and membrane bound immunoglobulin μ heavy chains is directed by mRNAs that differ at their 3′ ends. Cell 20:293–302.

Amara, S. G., Jonas, V., Rosenfeld, M. G., Ong, E. S., and Evans, R. M. (1982). Alternative RNA processing in calcitonin gene expression generates mRNAs encoding different polypeptide products. Nature 298:240–244.

Amara, S. G., Evans, R. M., and Rosenfeld, M. G. (1984). Calcitonin/CGRP transcription unit: tissue specific expression involves selective use of alternative polyadenylation sites. Molec. Cell Biol. 4:2151–2160.

Andreadis, A., Gallego, M., and Nadal-Ginard, B. (1987). Generation of protein isoform diversity by alternative splicing. Ann. Rev. Cell Biol. 3:207–242.

Birnstiel, M. L., Busslinger, M., and Strub, K. (1985). Transcription termination and 3' end processing: the end is in site! Cell 41:349–359.

Boggs, R. T., Gregor, P., Idriss, S., Belote, J. M., and McKeown, M. (1987). Regulation of sexual differentiation in D. melanogaster via alternative splicing of RNA from the transformer gene. Cell 50:739–747.

Bovenberg, R. A. L., van de Meerendonk, W. P. M., Baas, P. D., Steenbergh, P. H., Lips, C. J. M., and Jansz, H. S. (1986). Model for alternative splicing in human calcitonin gene expression. Nucleic Acids Res. 14:8785–8803.

Brandt, P. W., Diamond, M. S., and Schachat, F. H. (1984). The thin filament of vertebrate skeletal muscle co-operatively activates as a unit. J. Molec. Biol. 180:379–384.

Brawerman, G. (1987). Determinants of messenger RNA stability. Cell 48:5–6.

Breitbart, R. D., and Nadal-Ginard, B. (1986). Complete nucleotide sequence of the fast skeletal troponin-T gene. J. Molec. Biol. 188:313–324.

Breitbart, R. E., and Nadal-Ginard, B. (1987). Developmentally induced, muscle specific trans factors control the differential splicing of alternative and constitutive troponin-T exons. Cell 49:793–803.

Breitbart, R. E., Nguyen, H., Medford, R., Destree, A., Mahdavi, V., and Nadal-Ginard, B. (1985). Intricate combinatorial patterns of exon splicing generate multiple regulated troponin-T isoforms from a single gene. Cell 41:67–82.

Breitbart, R. E., Andreadis, A., and Nadal-Ginard, B. (1987). Alternative splicing; a ubiquitous mechanism for the generation of multiple protein isoforms from single genes. Ann. Rev. Biochem. 56:467–495.

Briggs, M. M., Klevit, R. E., and Schachat, F. H. (1984). Heterogeneity of contractile proteins. J. Biol. Chem. 259:10369–10375.

Briggs, M. M. Lin, J. J.-C., and Schachat, F. H. (1987). The extent of amino terminal heterogeneity in rabbit fast skeletal muscle troponin-T. J. Muscle Res. Cell Motil. 8:1–12.

Buckingham, M. E., and Minty, A. J. (1983). Contractile protein genes. In Eukaryotic Genes: Their Structure, Activity, and Regulation. Maclean, N., Gregory, S. P., Flavell, R. A., eds. Butterworths, London, pp. 365–395.

Caras, I. W., Davitz, M. A., Rhee, L., Weddell, G., Martin, D. W., and Nussenzweig, V. (1987). Cloning of decay-accelerating factor suggests novel use of splicing to generate two proteins. Nature 325:545–549.

Cech, T. (1986). The generality of self-splicing RNA: relationship to nuclear mRNA splicing. Cell 44:207–210.

Cech, T., and Bass B. (1986). Biological catalysis by RNA. Ann. Rev. Biochem. 55:599–629.

Chu, F., Maley, G., Maley, F., and Belfort, M. (1984). Intervening sequence in the thymidylate synthase gene of bacteriophage T4. Proc. Natl. Acad. Sci. USA 81:3049–3053.

Chu, F., Maley, G., West, D., Belfort, M., and Maley, F. (1986). Characterization of the intron in the phage T4 thymidylate synthase gene and evidence for its self-excision from the primary transcript. Cell 45:157–166.

Collins, T., Bonthron, D. T., and Orkin, S. H. (1987). Alternative RNA splicing affects function of encoded platelet derived growth factor A-chain. Nature 328:621–624.

Cory, S., and Adams, J. M. (1980). Deletions are associated with somatic rearrangement of immunoglobulin heavy chain genes. Cell 19:37–51.

Craik, C. S., Sprang, S., Fletterick, R., and Rutter, W. J. (1982). Intron–exon splice junctions map at protein surfaces. Nature 299:180–182.

Crenshaw, E. B., Russo, A. F., Swanson, L. W., and Rosenfeld, M. G. (1987). Neuron specific alternative RNA processing in transgenic mice expressing a methallothionein-calcitonin fusion gene. Cell 49:389–398.

Crick, F. H. C. (1953). The fourier transform of a coiled coil. Acta Cryst. 6:685–689.

Cunningham, B. A., Hemperly, J. J., Murray, B. A., Prediger, E. A., Brackenbury, R., and Edelman, G. M. (1987). Neural cell adhesion molecule: structure, immunoglobulin like domains, cell surface modulation and alternative RNA splicing. Science 236:799–806.

Danner, D., and Leder, P. (1985). Role of an RNA cleavage/poly(A) addition site in the production of membrane bound and secreted IgM mRNAs, Proc. Natl. Acad. Sci. USA 82:8658–8662.

Darnell, J. E. (1978). Implications of RNA-RNA splicing in the evolution of eukaryotic cells. Science 202:1257–1260.

Dickson, G., Gower, H. J., Barton, C. H., Prentice, H. M., Elsom, V. L., Moore, S. E., Cox, R. D., Quinn, C., Putt, W. and Walsh, F. S. (1987). Human muscle neural cell adhesion molecule (N-CAM); identification of a muscle specific sequence in the extracellular domain. Cell 50:1119–1130.

Early, P., Rogers, J., Davis, M., Calame, K., Bond, M., Wall, R., and Hood, L. (1980). Two mRNAs can be produced from a single immunoglobulin μ gene by alternative RNA processing pathways. Cell 20:313–310.

Eperon, L. P., Estibeiro, J. P., and Eperon, I. C. (1986). The role of nucleotide sequences in splice site selection in eukaryotic pre-messenger RNA. Nature 324:280–282.

de Ferra, F., Engh, H., Hudson, L., Kamholz, J., Puckett, C., Molineaux, S., and Lazzarini, R. A. (1985). Alternative splicing accounts for the four forms of myelin basic protein. Cell 43:721–727.

Fink, G. R. (1986). Translational control of transcription in eukaryotes. Cell 45:155–156.

Flicker, P. F., Philips, G. N., and Cohen, C. (1982) Troponin and its interactions with tropomyosin. J. Molec. Biol. 162:495–501.

Fontaine, B., Klarsfeld, A., and Changeux, J. P. (1987). Calcitonin gene related peptide and muscle activity regulate acetylcholine receptor α-subunit mRNA levels by distinct intracellular pathways. J. Cell Biol. 105:1337–1342.

Freytag, S. O., Beaudet, A. L., Bock, H. G. O., and O'Brien, W. E. (1984). Molecular structure of the human argininosuccinate synthetase gene: occurrence of alternative mRNA splicing. Molec. Cell Biol. 4:1978–1974.

Galli, G., Guise, J. W., McDevitt, M. A., Tucker, P. W. and Nevins, J. R. (1987). Relative position and strength of poly(A) sites as well as transcription termination are critical to membrane versus secreted μ-chain expression during B-cell development. Genes and Development 1:471–481.

Gil, A., and Proudfoot, N. J. (1987). Position dependent sequence elements downstream of AAUAAA are required for efficient rabbit β-globin mRNA 3' end formation. Cell 49:399–406.

Gil, G., Smith, J. R., Goldstein, J. L., and Brown, M. S. (1987). Optional exon in the 5' untranslated region of 3-hydroxy-3-methylglutaryl coenzyme A synthase gene: conserved sequence and splicing pattern in humans and hamsters. Proc. Natl. Acad. Sci. USA 84:1863–1866.

Gilbert W. (1978). Why genes in pieces? Nature 271:501

Glaichenhaus, N., and Cuzin, F. (1987). A role for ID repetitive sequences in growth and transformation dependent regulation of gene expression in rat fibroblasts. Cell 50:1081–1089.

Golub, E. S. (1987). Somatic mutation: diversity and regulation of the immune repertoire. Cell 48:723–724.

von Heijne, G. (1986). Mitochondrial targetting sequences may form amphiphilic helices. EMBO J. 5:1335–1342.

Helfman, D., Cheley, S., Kuismanen, E., Finn, L. A., and Yamawaki-Kataoka, Y. (1986). Nonmuscle and muscle tropomyosin isoforms are expressed from a single gene by alternative RNA splicing and polyadenylation. Molec. Cell Biol. 6, 3582–3595.

Homcy, C. J., Scott, B., Sullivan, K., and Nadal-Ginard, B. Generation of multiple $G_{s\alpha}$ mRNAs by alternative splicing: identification of a novel isoform in canine ventricle. J. Biol. Chem. (submitted).

Humphries, M. J., Akiyama, S. K., Komoriya, A., Olden, K., and Yamada K. M. (1986). Identification of an alternatively spliced site in human plasma fibronectin that mediates cell type specific adhesion. J. Cell Biol. 103:2637–2647.

Hunkapillar, T., Huang, H., Hood, L., and Campbell, J. H. (1982). The impact of modern genetics on evolutionary theory. In Perspectives on Evolution. R. Milkman, ed Sinauer Associates, Sunderland, Mass., pp. 170–172.

Hynes, R. (1985). Molecular biology of fibronectin. Ann. Rev. Cell Biol. 1:67–90.

Imamura, K. and Tanaka, T. (1982). Pyruvate kinase isozymes from rat. Meth. Enzymo. 90:150–165.

Irminger, J. C., Rosen, K. M., Humbel, R. E., and Villa-Komaroff, L. (1987). Tissue specific expression of insulin like growth factor II mRNAs with distinct 5' untranslated regions. Proc. Natl. Acad. Sci. USA 84:6330–6334.

Izumo, S., and Mahdavi, V. (1988). Thyroid hormone receptor α isoforms generated by alternative splicing differentially activate myosin HC gene transcription. Nature 334:539–542.

Kaine, B., Gupta. R., and Woese, C. (1983). Putative introns in tRNA genes of prokaryotes. Proc Natl. Acad. Sci USA 80:3309–3312.

Kelley, D. E., and Perry, R. P. (1986). Transcriptional and posttranscriptional control of immunoglobulin mRNA production during B lymphocyte development. Nucleic Acids Res. 14:5431–5447.

Kornblihtt, A. R., Vibe-Pedersen, K., and Baralle, F. E. (1984). Human fibronectin: cell specific alternative mRNA splicing generates polypeptide chains differing in the number of internal repeats. Nucleic Acids Res. 12:5853–5868.

Kornblihtt, A. R., Umezawa, K., Vibe-Pedersen, K., and Baralle, F. E. (1985). Primary structure of human fibronectin: differential splicing may generate at least 10 polypeptides from a single gene. EMBO J. 4:1755–1759.

Konarska, M. M., and Sharp, P. A. (1987). Interactions between small nuclear ribonucleoprotein particles in formation of spliceosomes. Cell 49:763–774.

Kozak, J. (1986). Regulation of protein synthesis in virus infected animal cells. Adv. Virus Res. 31:229–292.

Kwiatkowski, D. J., Stossel, T. P., Orkin, S. H., Mole, J. E., Colten, H. R., and Yin, H. L. (1986). Plasma and cytoplasmic gelsolins are encoded by a single gene and contain a duplicated actin binding domain. Nature 323:455–458.

Ladner, M. B., Martin, G. A., Noble, J. A., Nikoloff, D. M., Tal, R., Kawasaki, E. S., and White, T. J. (1987). Human CSF-1: gene structure and alternative splicing of mRNA precursors. EMBO J. 6:2693–2698.

Lang, K. M., and Spritz, R. A. (1987). In vitro splicing pathways of pre-mRNAs containing multiple intervening sequences. Molec. Cell Biol. 7:3428–3437.

Laski, F. A., Rio, D. C., and Rubin, G. M. (1986). Tissue specificity of Drosophila P element transposition is regulated at the level of mRNA splicing. Cell 44:7–19.

Leff, S. E., Evans, R. M., and Rosenfeld, M. G. (1987). Splice commitment dictates neuron specific alternative RNA processing in calcitonin/CGRP gene expression. Cell 48:517–524.

Lillie, J. W., Green, M., and Green, M. R. (1986). An adenovirus Ela protein region required for transformation and transcriptional repression. Cell 46:1043–1051.

Lillie, J. W., Loewenstein, P. M., Green, M. R., and Green, M. (1987). Functional domains of adenovirus type 5 Ela proteins. Cell 50:1091–1100.

Lone, Y. C., Simon, M.-P., Kahn, A., and Marie, J. (1986). Sequences complementary to the brain specific "identifier" sequences exist in L-type pyruvate kinase mRNA (a liver specific messenger) and in transcripts especially abundant in muscle. J. Biol. Chem. 261:1499–1502.

Lund, E., Kahan, B., and Dahlberg, J. E. (1985). Differential control of U1 small nuclear RNA expression during mouse development. Science 229:1271–1274.

MacLeod, A. R., Houlker, C. Reinach, F. E., Smillie, L. B., Talbot, K., Modi G., and Walsh, F. S. (1985). A muscle type tropomyosin in human fibroblasts: evidence for expression by an alternative RNA splicing mechanism. Proc. Natl. Acad. Sci. USA 82:7835–7839.

Maeda, N., and Smithies, O. (1986). The evolution of multigene families: human haptoglobin genes. Ann. Rev. Genet. 20:81–108.

Maniatis, T., and Reed, R. (1987). The role of small nuclear ribonucleoprotein particles in pre-mRNA splicing. Nature 325:673–678.

Mardon, H. J., Sebastio, G., and Baralle, F. E. (1987). A role for exon sequences in alternative splicing of the human fibronectin gene. Nucleic Acids Res. 15:7725–7733.

Medford, R. M., Nguyen, H., Destree, A., Summers, E., and Nadal-Ginard, B. (1984). A novel mechanism of alternative RNA splicing for the developmentally regulated generation of troponin-T isoforms from a single gene. Cell 38:409–421.

Mitchell, P. J., Urlaub, G., and Chasin, L. (1986). Spontaneous splicing mutations at the dihydrofolate reductase locus in chinese hamster ovary cells. Molec. Cell Biol. 6:1926–1935.

Mowry, K. L., and Steitz, J. A. (1987). Identification of the human U7 snRNP as one of several factors involved in the 3' end maturation of histone premessenger RNA's. Science 238:1682–1687.

Mueller, P. P., and Hinnebusch, A. G. (1986). Multiple upstream AUG codons mediate translational control of GCN4. Cell 45:201–207.

Muirhead, H., Clayden, D. A, Barford, D., Lorimer, C. G., Fothergill-Gilmore, L. A., Sciltz, E., and Schmitt, W. (1986). The structure of cat muscle pyruvate kinase. EMBO J. 5:475–481.

Nabeshima, Y., Fujii-Kuriyama, Y., Muramatsu, M. and Ogata, K. (1984). Alternative transcription and two modes of splicing result in two myosin light chains from one gene. Nature 308:333–338.

Noguchi, T., Inoue, H., and Tanaka, T. (1986). The M1 and M2 type isozymes of rat pyruvate kinase are produced from the same gene by alternative RNA splicing. J. Biol. Chem. 261:13807–13812.

Ohno, S. (1970). Evolution by Gene Duplication. Springer-Verlag, New York.

Okazaki, K., Davis, D., and Sakano, H. (1987). T cell receptor β gene sequences in the circular DNA of thymocyte nuclei: direct evidence for intramolecular DNA deletion in V-D-J joining. Cell 49:477–485.

Padgett, R. A., Grobowski, P. J., Konarska, M. M., Seiler, S., and Sharp, P. A. (1986). Splicing of messenger RNA precursors. Ann. Rev. Biochem. 55:1119–1150.

Pearlstone, J. R., and Smillie, L. B. (1983). Effects of troponin-I plus -C on the binding of Troponin-T and its fragments to α-tropomyosin. J. Biol. Chem. 258:2534–2542.

Pearlstone, J. R., Carpenter, M. R., Johnson, P., and Smillie, L. B. (1977). Primary structure of rabbit muscle troponin-T. J. Biol. Chem. 252:983–989.

Periasamy, M., Strehler, E., Garfinkel, L. I., Gubits, R. M., Ruiz-Opazo, N., and Nadal-Ginard, B. (1984). Fast muscle myosin light chains 1 and 3 are produced from a single gene by a combined process of differential RNA transcription and splicing. J. Biol. Chem. 259:13595–13604.

Peterson, M. L., and Perry, R. P. (1986). Regulated production of μ_m and μ_s mRNA requires linkage of the poly(A) addition sites and is dependent on the length of the μ_s-μ_m intron. Proc. Natl. Acad. Sci. USA 83:8883–8887.

Pfeifer, M., Karch, F., and Bender, W. (1987). The bithorax complex: control of segmental identity. Genes and Development 1:891–898.

Platt, T. (1986). Transcription termination and the regulation of gene expression. Ann. Rev. Biochem. 55:339–372.

Proudfoot, N., and Brownlee, G. G. (1976). 3' non-coding sequences in eukaryotic messenger RNA Nature 263:211–214.

Raghow, R. (1987). Regulation of messenger RNA turnover in eukaryotes. Trends Biochem. Sci. 12:122–166.

Reed, R., and Maniatis, T. (1985). Intron sequences involved in lariat formation during pre-mRNA splicing. Cell 41:95–105.

Reed, R., and Maniatis, T. (1986). A role for exon sequences and splice site proximity in splice site selection. Cell 46:681–690.

Reuther, J. E., Maderious, A., Lavery, D., Logan, J., Man Fu, S., and Chen-Kiang, S. (1986). Cell type specific synthesis of murine immunoglobulin μ RNA from an adenovirus vector. Molec. Cell Biol. 6:123–133.

Reynolds, G. A., Goldstein, J. L., and Brown, M. S. (1985). Multiple mRNAs for 3-hydroxy-3-methylglutaryl coenzyme A reductase determined my multiple transcription initiation sites and intron splicing sites in the 5'-untranslated region. J. Biol. Chem. 260:10369–10377.

Rio, D. C., Laski, F. A., and Rubin, G. M. (1986). Identification and immunochemical analysis of biologically active Drosophila P element transposase. Cell 44:21–32.

Rogers, J., Early, P., Carter, C., Calame, K., Bond, M., Hood, L., and Wall, R. (1980). Two mRNAs with different 3' ends encode membrane bound and secreted forms of immunoglobulin μ chain. Cell 20:303–312.

Rogers, S. L., Letourneau, P. C., Peterson, B. A., Furcht, L. T., and McCarthy, J. B.

(1987). Selective interaction of peripheral and central nervous system cells with two distinct cell binding domains of fibronectin. J. Cell Biol. 105:1435–1442.

Roise, D., Horvath, S. J., Tomich, J. M., Richards, J. H., and Schatz, G. (19877). A chemically synthesized pre-sequence of an imported mitochondrial protein can from an amphiphilic helix and perturb natural and artificial phospholipid bilayers. EMBO J. 5:1327–1334.

Romans, P., Firtel, R. A., and Saxe, C. L. (1985). Gene specific expression of the actin multigene family Dictyostelium discoideum. J. Molec. Biol. 186:337–355.

Rosenfeld, M. G., Amara, S. G., and Evans, R. M. (1984) Alternative RNA processing: determining neuronal phenotype. Science 225:1315–1320.

Ruiz-Opazo, N., Weinberger, J., and Nadal-Ginard, B. (1985). Comparison of α-tropomyosin sequences from smooth and striated muscle. Nature 5:67–70.

Ruiz-Opazo, N., and Nadal-Ginard, B. (1987). α-Tropomyosin gene organization. J. Biol. Chem. 261:4755–4765.

Sakano, H., Huppi, K., Heinrich, G., and Tonegawa, S. (1979) Sequences at the somatic recombination sites of immunoglobulin light-chain genes. Nature 280:288–294.

Sakano, H., Kurosawa, Y., Weigert, M., and Tonegawa, S. (1981). Identification and nucleotide sequence of a diversity DNA segment (D) of immunoglobulin heavy-chain genes. Nature 290:562–565.

Salditt-Georgieff, M., Harpold, M., and Chen-Kiang, S. (1980). The addition of 5' cap structures occurs early in hnRNA synthesis and prematurely terminated molecules are capped. Cell 19:69–78.

Schachat, F. H., Diamond, M. S., and Brandt, P. W. (1987). The effect of different troponin T-tropomyosin combinations on thin filament activation. J. Mol. Biol. 198:551–554.

Schwarzbauer, J. E., Patel, R. S., Fonda, D., and Hynes, R. O. (1987). Multiple sites of alternative splicing of the rat fibronectin gene transcript. EMBO J. 6:2573–2580.

Seidman, J. G., Nau, M. M., Norman, B., Kwan, S. P., Scharff, M., and Leder, P. (1980). Immunoglobulin V/J recombination is accompanied by deletion of joining site and variable region segments. Proc. Natl. Acad. Sci. USA 77:6022–6026.

Sharp, P. (1987). Splicing of messenger RNA precursors. Science 235:766–771.

Shaw, G., and Kamen, R. (1986) A conserved AU sequence from the 3' untranslated region of GM-CSF mRNA mediates selective mRNA degradation. Cell 46:659–667.

Skolnick-David, H., Moore, C. L., and Sharp, P. A. (1987). Electrophoretic separation of poladenylation-specific complexes. Genes and Development 1:672–682.

Smith, C. W. J., and Nadal-Ginard, B. (1989). Mutually exclusive splicing of α-tropomyosin exons enforced by an unusual lariat branch point location: implications for constitutive splicing. Cell 56:749–758.

Smith, C. W. J., Patton, J. G., and Nadal-Ginard, B. (1989). Alternative splicing in the control of gene expression. Ann. Rev. Genetics 23: in press.

Solnick, D. (1985). Alternative RNA splicing caused by RNA secondary structure. Cell 43:667–676.

Solnick, D., and Lee, S. I. (1987). Amount of RNA secondary structure required to induce an alternative splice. Molec. Cell Biol. 7:3194–3198.

Strehler, E. E., Periasamy, M., Strehler-Page, M.-E. and Nadal-Ginard, B. (1985). Myosin light chain 1 and 3 gene has two structurally distinct and differentially regulated promoters evolving at different rates. Molec. Cell Biol. 5:3168–3182.

Stuart, D. I., Levine, M., Muirhead, H., and Stammers, D. K. (1979). Crystal structure of cat muscle pyruvate kinase at a resolution of 2.6 A. J. Mol. Biol. 134:109–142.

Tong, B. T., Auer, D. E., Jaye, M.., Kaplow, J. M., Ricca, G., McConathy, E., Drohan, W., and Deuel, T. F. (1987). cDNA clones reveal differences between human glial and endothelial cell platelet derived growth factor A-chains. Nature 328:619–621.

Tsurushita, N., Avdalovic, N. M., and Korn, L. J. (1987). Regulation of differential processing of mouse immunoglobulin μ heavy chain mRNA. Nucleic Acids Res. 15:4603–4615.

Tzamarias, D., Alexandraki, D. and Thireos, G. (1986). Multiple *cis*-acting elements modulate the translational efficiency of GCN4 mRNA in yeast. Proc. Natl. Acad. Sci. USA 83:4849–4853.

Wieczorek, D., Smith, C. W. J., and Nadal-Ginard, B. (1988). The rat α-tropomyosin gene generates a minimum of six different mRNAs coding for striated, smooth and non-muscle isoforms by alternative splicing. Mol. Cell Biol. 8:679–694.

Wolf, O., Kourides, I. A., and Gurr, J. A. (1987) Expression of the gene for the β subunit of mouse thyrotropin results in multiple mRNAs differing in their 5' untranslated regions. J. Biol. Chem. 262:16596–16603.

Young, R. A., Hagenbuchle, O., and Schibler, U. (1981). A single mouse α-amylase gene specifies two different tissue specific mRNAs. Cell 23:451–458.

Yuan, D., and Tucker, P. W. (1984). Transcriptional regulation of the μ-δ heavy chain locus in normal murine B lymphocytes. J. Exp. Med. 160:564–583.

Zarkower, D., and Wickens, M. (1987). Formation of mRNA 3' termini: stability and dissociation of a complex involving the AAUAAA sequence. EMBO J. 6:177–186.

Zeevi, M., Nevins, J. R., and Darnell, J. E. (1981). Nuclear RNA is spliced in the absence of poly(A) addition. Cell 26:39–46.

Zhuang, Y., Leung, H. and Weiner, A. M. (1987). The natural 5' splice site of simian virus 40 large T antigen can be improved by increasing the base complementarity to U1 RNA. Molec. Cell Biol. 7:3018–3020.

Ziff, E. B. (1982). Transcription and RNA processing by the DNA tumour viruses. Nature 287:491–499.

Index

Actin, 27, 48
Active site, enzymes, 11
Adenovirus
 E1a protein, 178
 mRNA, 92–93
Albumin/α-fetoprotein, 106
Alcohol dehydrogenase, 22, 23, 29, 81, 86
 exon shuffling, 108
 exon sizes, 84
 liver, 4, 76–79, 83
 maize, 77–79
 NAD-binding domain, 78, 79, 86
Alternative splicing. *See* Splicing, alternative, mRNA
Archaebacteria, introns, 119, 120, 122, 151
Argininosuccinate synthetase, 182
Aspartate carbamoyltransferase, 22
Aspergillus, 47, 150
 nidulans, 29, 150
 triose phosphate isomerase, 107
ATP
 binding, exons, 22, 24
 splicing mechanisms, introns, 133, 135, 137, 139

Bacteriophage T4, 127, 152
 RNA processing, 30

Caenorhabditis elegans, 113, 136
Calcitonin gene-related peptide, 168, 169, 172, 181
Calcium, 180
Calcium-binding domain, exon shuffling, 15, 19–21, 108
Capping, RNA processing, 163
α-Carbon distance plots, secondary structures, 70–75, 77–79
 GADPH, 78–81
 LDH, 73–75, 77, 81
Carboxyl terminals
 α-β unit secondary structures, 80, 82–83
 cf. N-terminals, secondary structures, 76–77

Cassette exons, 166, 167, 173, 175, 177, 179
Catalase, fungal, 7
CBP2 gene, yeast, splicing mechanisms, 143
Cell adhesion molecules, neural, 179
Chlamydomonas, 123, 149, 151
Chloroplast DNA, 117, 119, 120, 122
cis-acting elements, alternative mRNA splicing, 162, 172–174
Coagulation and fibrinolytic proteins, 13–21
cob gene, *Neurospora crassa*, 141–143
Collagen, 27, 28, 106
Colony-stimulating factor-1, 183
Compact modules, protein structure, 24
Complement
 C3b, 177
 C4b, 177
 C9, 14, 17, 18, 20, 46, 65, 109
 factor B, 24
Conalbumin, 34
Constitutive mRNA splicing, cf. alternative splicing, 162, 165, 186
Contour cells, definition, 73
coxI gene, *Saccharomyces*, 149, 150, 151
coxII gene, mitochondrial, 150
Cryptic splice sites, alternative mRNA splicing, 165
β-Crystallins, 34
C-value, intron, 57, 59
Cyclization, splicing mechanisms, introns, 140–142
Cytochrome *b*, 5, 6, 26, 123
Cytochrome *c*, 4, 5, 26

Decay accelerating factor, 176–178
Default splicing pattern, mRNA, 167–168, 171, 174
Dehydrogenases, NAD-binding domain, 4–7, 11, 24, 32; *see also specific enzymes*
Dictyostelium discoideum, 119
Distance plots, α-carbon intron-structure relationships, 70–75, 77–79

Distance plots (*continued*)
  GAPDH, 78–81
  LDH, 73, 74, 75, 77, 81
DNA
  exon shuffling, 43
  introns, evolutionary age, 32–34
Domains, 11
  active site, enzymes, 11
  binding function, 11
    protease, 21
  calcium-binding, exon shuffling, 15, 19–21, 108
    Kringle, 15, 16–21
    plasma protein, 18
    protease, 21
  definition, 6–7, 68–69, 73
  existence before DNA coding, 88, 89
  exon-shuffling, 12–13, 45
  fibronectin, types I and II, 15, 17, 18, 20
  heme-binding, 7, 25, 26
  history of concept, 4–7
  and introns, relationship (Blake), 64–65, 68–69, 73, 84, 86
  NAD-binding, dehydrogenases, 4–7, 11, 24, 32, 76–80, 87, 88
  ovoinhibitor gene introns
    alignment, 103
    coding sequences, duplication, 104–106
  plasma proteins, homology, 18
  structure, ovomucoid gene, 95–99, 105
Drosophila, 122
  homeobox, 182
  melanogaster, 113, 119
  P-element transposase, 178
  28S rRNA genes, 92
  tra gene, 175, 178
Duchenne muscular dystrophy gene, 113
Duplication, structural genes, 10–12, 26

Epidermal growth factor, 14–18, 21, 46, 65, 108–109
  A units, 14, 15, 18, 20
  B units, 14, 15, 18
  precursor, 14–16, 109
Escherichia coli, 43, 51, 107
  triose phosphate isomerase, 107
Eukaryotes, genes, cf. prokaryotes, 10, 11, 12
  energy expenditure hypothesis, 33
  introns, 44, 50–51, 57
    mitochondria, 50
    plastids, 50
  cf. progenotes, 63–65
    evolutionary schemes, 66

Evolution
  collagen, 27, 28
  engineering vs. tinkering (Jacob), 52
  exon shuffling, 19, 108–109
    as "edge," 53
  gene splicing errors and, 81
  of globins, 25–26
  prokaryotic, 25
  as history vs. process, 43, 51
  introns, 87
    evolutionary age, 26–29
    evolution of genes, 82–83
    neo-Darwinian view of, 52–56, 59
    see also Intron function and evolution
  introns-early vs. introns-late view, 44, 45, 47, 49, 50–51, 52
    progenotic genes, 63–65, 83
  metaphors for, 51–53
  of ovomucoid gene, 99–102, 104–106
  pathway, exons, 23
  progenotic genes
    intron-dependent evolution, 82–83
    "introns-early" view, 63–65, 83
  puncutated equilibria, 56, 59
  rate of, point mutation and, 24, 25
  RNA processing, 29–30
    alternative, 183–188
  split genes, 21–22, 26
Exaptation, intron function and evolution, 54
Exon(s)
  cassette, 166, 167, 173, 175, 177, 179
  domain concept, history of, 7
  glycolytic enzymes, ancient, 22–24
    sizes of exons, 84, 86
    see also specific enzymes
  mosaic gene evolution, 33–35
  multiple duplication, 26
  secondary structure, 35
  terminal 3' noncoding sequence, 85
Exon shuffling, 10–13, 43–47, 58, 78–79, 187
  ADH, 108
  DNA cf. RNA level, 43
  domains, 12–13, 45
  and evolution, 19, 53, 108–109
  Gilbert-Blake hypothesis, 12–13, 16, 43, 45, 49, 64
  protein and gene structure correlations, 45–46
    blood coagulation and fibrinolytic proteins, 13–21
    glycolytic genes, ancient, 46–47, 108–109
  new genes, 46
  supersecondary units, 45, 47

INDEX

Factor Va, 20
Factor IX, 14, 15, 16, 19, 20, 21, 46, 65, 108, 109
Factor X, 14, 16, 19, 20, 21, 46, 109
Factor XII, 14, 15, 16, 18–21
α-Fetoprotein, 26, 106
Fibrinogen genes, 48
Fibronectin, 21, 26, 35, 108, 175
    type I and II domains, 15, 17, 18, 20
    type III unit, 20
Flavodoxin, 5–7

Gelsolin, 177
Gene(s)
    fusion mechanism, 12
    splicing errors, evolutionary effects, 81
    see also Eukaryotes, genes, cf. prokaryotes;
        Split genes; *specific genes*
Globin genes
    β-, 73
        split genes, rabbit, 93
    evolution, 25–26
    prokaryotic, 25
Glucagon, 34
Glyceraldehyde-3-phosphate dehydrogenase (GAPDH), 4, 6, 47, 65, 67, 68, 83, 86, 88
    distance plots, intron-structure relationship, 78–81
    NAD binding, 78–79, 80, 86
    exon shuffling, 22, 23, 108
    exon sizes, 84
Glycine max, 119
Glycolytic enzymes
    ancient, exon shuffling, 22, 46–47
    progenotic genes as, 65–68
    see also specific glycolytic enzymes
GTP, splicing mechanisms, introns, 141, 142
GTP-binding protein, Gs α, 184

Hairpin structures, RNA processing, 173
Halobacterium
    mediterranei, 139
    volcanii, 139
Heme-binding domain, 7, 25, 26
Hemoglobin, 5, 6
    myoglobin homology with, 3–4
    normal cf. sickle-cell, 3
    structure, diagram, 25
Hierarchies, intron function and evolution, 54–56
HMG CoA reductase, 182
HMG CoA synthase, 182, 183

Hydrogen bonding patterns, lactate dehydrogenase, 75

Immunoglobulin genes, 184
μ_m and μ_s, 168–170, 176
    history of domain concept, 4
Insulin-like growth factor II, 182
Interleukin-2, 46
Internal guide sequence, introns, 125, 126, 142, 145
Intron(s), 12
    discovery, 43, 92–93, 112
    domain concept, history of, 7
    and domain relationships (Blake), 64–65, 68–69, 73, 84, 86
    early hypotheses regarding, 93–94
    evolution
        age, 26–29
        classification into two groups, 87
        DNA, 32–34
        early origins, 106–108
        interspecies comparisons, 28
        introns-early vs. introns-late view, 44, 45, 47, 49, 50–51, 52, 63–65, 83
        static vs. dynamic viewpoint, 93
        stability of patterns, 27–28
        see also Intron function and evolution
    intragenic duplication, 106
    locations, 23
    loss/gain of, 16, 44, 47–50, 86–87
        phylogenetic analyses, 48
    as "mutational sinks," 95
    optional components of genes, 148–151
        mobility or infectivity, 151
    ovoinhibitor gene, 102–106
    patterns
        probabilities of chance arrangements, 82–84
        protein structure constraints, nonrandomness, 84
    prokaryote-eukaryote relationship, 44, 50–51, 57
    mitochondria, 50
    plastids, 50
    and RNA processing, 63–66, 77, 112
        evolutionary age, 29–32
    "selfish," 44, 49–50
        mitochondrial data, 49
    and structure relationship, GADPH, 78–81
    see also Splicing mechanisms, introns
Intron classification, 113–130
    group I introns, 123–128
        conserved RNA secondary structures, 125
        internal guide sequence, 125, 126
        as mobile genetic elements, 148–153

Intron classification (*cont.*)
  group I introns (*cont.*)
    ORFs, 126–128
  group II introns, 123, 124, 128–130
    conserved core secondary structure, 129, 130
    as mobile genetic elements, 148–153
    ORFs, 129, 152
    reverse transcription, 152–153
    splicing mechanisms, 144–148
  nuclear pre-mRNA introns, 113–117, 123, 124, 129
    boundaries of, conserved, 114–115
    5' end consensus sequence, 116
    interspecies splicing, 116
    splicing mechanisms, 130–136
  tRNA gene intron, 117–123, 151
    compilation of DNA sequences for, various species, 118–119
    secondary structure diagram, 120
    splicing mechanism, 136–139
    stem-loop structures, 121
Intron function and evolution, 44, 51–60
  adaptation, 54–56, 58–59
  constraints, 57–58
  exaptation, 54
  hierarchies, 54–56
  molecular biologists' view of evolution, 52
  neo-Darwinian view of evolution, 52, 53, 54–56, 59
  as remnants of genome assembly, precellular evolution, 52
  schema, 53
  selection against, 56–57
    DNA content (C-value), 57, 59
  selection for, 54–56, 58–59
  speciation, 55–56, 58–59

Kazal serine proteinase inhibitors, 102, 104, 105
  tertiary structure, 104
Kluyveromyces, 150
  thermotolerans, 150
Kringle domain, exon shuffling in plasma proteins, 15, 16–21

Lactate dehydrogenase, 5, 6, 22, 86
  distance plot patterns, 73, 74, 75, 77, 81
  hydrogen bonding patterns, 75
  NAD binding region, 76
  nicotinamide binding region, 76
  exon sizes, 84
  isozymes, 4
Lariat structures, RNA processing, 131–133, 144–148, 163, 165, 172

LDL receptor, 14–16, 46, 65, 108–109
Leghemoglobin, 27
Lipoprotein, low-density, receptor, 14–16, 46, 65, 108–109
Liver, alcohol dehydrogenase, 4, 76–79, 79, 83
Lobe, definition, 69, 87–88
Loop structures, RNA processing, 173

Maize alcohol dehydrogenase, 77–79
Malate dehydrogenase, 22, 76, 77, 78, 81
α-Melanocyte stimulating hormone, 34
Mitochondria
  coxII gene, 150
  introns
    fungal, 128
    prokaryote-eukaryote relationship, 50
    selfish, 49
    yeast, 123, 127, 128
Mobile genetic elements, 148–153
Modules, 24, 73
Mosaic genes, 10
  evolution, 33–35
  origins, 26–35
α-MSH, 34
Muscular dystrophy, Duchenne, gene, 113
Mutational sinks, introns as, 95
Myelin basic protein, 175
Myoglobin, 3–4, 25
Myosin-heavy chain genes, rat cf. nematode, 28–29, 34
Myosin light-chain gene, 168, 173, 174, 183, 187

NAD-binding domain, dehydrogenases, 4–7, 11, 24, 32, 76–80, 86–88
The Nature of Selection (Sober), 54–55, 56
Neo-Darwinian view of intron evolution, 52, 53, 54–56, 59
Neural cell adhesion molecules, 179
Neurospora, 17, 148, 151–153
  crassa, 119, 127, 128, 158
  cob gene, 141–143
  cyt-18 gene, 143–144
  RNA processing, evolution, 30
  intermedia, 150
Nicotiana rustica, 119
Nicotinamide binding region, lactate dehydrogenase, 76
Nuclear pre-mRNA introns, 113–117, 123, 124, 129
  splicing mechanisms, 130–136

Open reading frames, introns, 126–128, 129, 152

INDEX

Ovalbumin, 93–94
Ovalbumin-α-antitrypsin, 27
Ovoinhibitor gene introns, 102–106
  domains
    alignment, 103
    coding sequences, duplication, 104–106
    and Kazal serine proteinase inhibitors, 102, 104, 105
    tertiary structure, 104
  secondary structure, 104
  sequence organization, 105
Ovomucoid, 26
  domain structure, 95–99, 105
  evolution, 99–102, 104–106
    amino acid sequence, 97
    cf. mammalian trypsin inhibitors, 96–97, 100, 101
    schema, 101
    signal peptide, 100, 101
  mRNA, 95–96, 98, 99
  structural organization, 96

Phosphoglycerate kinase, 22–24, 29, 81
  exon shuffling, 108
  exon sites, 84
Physarum, 141, 149, 150
  polycephalum, 123
Plasma protein domains, homology, exon shuffling, 18
Plasminogen, 15, 16, 18–21
Platelet-derived growth factor, A-chain, 175, 177
Podospora, 148, 152
  anserina, 119
Point mutations
  and rate of evolution, 24, 25
  at splice junction, 64
Polyadenylation, RNA, 163, 164, 165
  alternative splicing, 168, 169, 170, 175
Preproinsulin genes, 28, 48
Pre-tRNA introns, 117–123, 151
  splicing mechanisms, 136–139
Primary structures, definition, 68
Processing. *See* RNA processing mechanisms; Splicing *entries*
Progenotes, 51
  definition, 63, 65
  energy-producing enzymes, existence prior to genes, 87–89
  cf. prokaryotes and eukaryotes, 63–65
    evolutionary schemes, 66
  RNA splicing, 63, 66
Progenotic genes
  evolution, intron facilitation, 82–83

evolution, "introns-early" view, 63, 64, 83
  protein structure, 64–65
  glycolytic enzymes as, 65–68; *see also specific glycolytic enzymes*
  unequal crossover mechanism, 12, 85, 89
Prokaryotes, globin gene evolution, 25; *see also* Eukaryotes, genes, cf. prokaryotes
Protease domain binding functions, exon shuffling, 21
Protein(s)
  diversity, alternative mRNA splicing, 183–185
  evolution, alternative mRNA splicing, 185–186
  and gene structure correlations, exon shuffling, 45–46
    blood coagulation and fibrinolytic proteins, 13–21
    glycolytic genes, ancient, 46–47, 108–109
    supersecondary units, 45, 47
  structural constraints
    alternative splicing, mRNA, 174–176
    α-helix and β-sheet, 176
    nonrandomness, intron patterns, 84
  structure, definitions, 24, 68–69
Protein C, 14, 15, 16, 19, 20, 21, 46, 109
Protein S, 15, 16, 20, 21
Protein Z, 15, 16, 21
Prothrombin, 15–21
Punctuated equilibria theory, 56, 59
Pyruvate kinase, 23, 32, 46–47, 81
  exon sizes, 84
  isoforms, 180, 183

Quaternary structure, definition, 69

Reverse transcription, introns, 152–153
RFLPs, 94
Rhodopsin, 34
Ribozyme, 31–32
RNA, messenger
  adenovirus, 92–93
  ovomucoid gene, 95–96, 98, 99
  RNA processing, evolution, 29, 30
  see also Splicing, alternative, mRNA
RNA, ribosomal, 51
  RNA processing, evolution, 30
  21S gene, 49
  28S genes, *Drosophila*, 92
  Tetrahymena, self-splicing, 50
RNA, transfer
  gene intron, 117–123, 151
  RNA processing, evolution, 29, 30

RNA (*continued*)
  synthesis, 5
  tRNA synthetase, 143, 144
RNA ligase, splicing mechanisms, introns, 137
RNA processing mechanisms, 163–166
  capping, 163
  and evolution, 29–30
  and exon shuffling, 43
  hairpin and loop structures, 173
  introns and, 63–66, 77, 112
    evolutionary age, 29–32
    progenotes, 63, 66
  lariat form of intron, 163, 165, 172
  polyadenylation, 163, 164, 165
  transesterification, 165
  see also Splicing, alternative, mRNA
RNPs
  small nuclear (snRNPs)
    alternative splicing, mRNA, 172
    splicing mechanisms, introns, 130, 135, 136, 148
  heteronuclear (hnRNPs), 135

Saccharomyces, 148, 149
  cerevisiae, 107, 113, 115–117, 119, 127, 149
  coxI gene, 149, 151
Schizosaccharomyces, 148
  pombe, 116, 119, 150
Seal cf. whale, myoglobin, 3
Secondary structures, 68–72
  α-carbon distance plots, 70–75, 77–79
  carboxyl terminals, α-β units, 80, 82–83
  C- cf. N-terminals, 76–77
  diagram, 120
  α-helix, 70, 71, 73, 75, 76, 79–80, 81, 176
  ovoinhibitor gene introns, 104
  β-sheet, 70, 71, 73, 75, 76, 79–82, 176
  β-strands, 22–23
Selection, intron function and evolution
  against, 56–57
    DNA content (C-value), 57, 59
  for, 54–56, 58–59
"Selfish" DNA, 49, 54
"Selfish" intron, 44, 49–50
  mitochondrial data, 49
Seminal fluid protein, 15, 21
Signal peptide, ovomucoid gene, 100, 101
Speciation, intron function and evolution, 55–56, 58–59
Spliced exon reopening, splicing mechanisms, introns, 145
Splice junction
  drift, 16, 27, 29
  point mutations, 64

Spliceosomes, 130
  assembly, 131, 133, 134–136
Splicing, alternative, mRNA, 162–163, 166–188
  alternative 3' end exons, 168–170
  cis-acting elements, 162, 172–174
  cf. constitutive mRNA splicing, 162, 165, 186
  cryptic splice sites, 165
  default splicing pattern, 167–168, 171, 174
  evolutionary aspects, 183–188
  functional effects on proteins, 176–182
  poly(A) sites, 168, 169, 170, 175
  protein structural constraints, 174–176
  trans-acting elements, 170–172
  transcripts with different 5' ends, 168
  cf. untranslated RNA regions, 182–183
  see also RNA processing mechanisms
Splicing mechanisms, introns, 130–148
  group I introns, 139–144
    cyclization, 140–142
    internal guide sequence, 142, 145
    proteins assist self-splicing in vitro, 143–144
  group II introns, 144–148
    lariats, 144, 145, 147, 148
    spliced exon reopening, 145
    transesterification, 144, 145, 148
    trans-splicing, 147, 148
    Y-shaped intron, 147
  nuclear pre-mRNA introns, 130–136
    lariat structures, 131, 132, 133
    snRNPs, 130, 135, 136
    transesterification, 131, 132, 133, 135
    whole-cell and nuclear extract systems for, 132–133
  pre-tRNA introns, 136–139
Split genes
  discovery, 92–94
  β-globin gene, rabbit, 93
  evolution, 21–22, 26
  unequal crossover, advantages of, 12
Static vs. dynamic viewpoint, introns, 93
Stem-loop structures, tRNA gene intron, 121
Structural genes, duplication, 10–12, 26
Supersecondary structures, 71–73, 76
  definition, 68
  exon shuffling, 45, 47
SV40 T antigen, 167, 173

T cell receptor genes, 184
Tertiary structure, definition, 69, 72
Tetrahymena, 149
  pyriformis, RNA processing, 30

self-splicing rRNA, 50
thermophila
    nuclear rDNA, 123
    rRNA, 125–126, 139, 141, 142, 144, 150
TF IIIA, 106
Thrombin, 108
Thyroglobulin, 26
Thyrotropin, β-subunit, 182
Tissue plasminogen activator, 14–21, 46, 108
trans-acting mechanisms
    alternative processing mRNA, 170–172
    intron splicing, 136, 147, 148
Transesterification, alternative mRNA splicing, 131, 132, 133, 135, 144, 145, 148, 165
Transforming growth factor, 14
Triose phosphate isomerase, 23, 29, 32
    Aspergillus nidulans, 107
    chicken, *E. coli* and yeast compared, 107
α-Tropomyosin, 168, 170–175, 179, 187–188
β-Tropomyosin, 168, 179
Troponin C, 179

Troponin I, 179
Troponin T, fast muscle, 167, 170, 172, 173, 175, 179–180, 187
    Ca^{2+} and, 180
Trypsin inhibitors, mammalian, cf. ovomucoid gene, 96–97, 100, 101

Unequal crossover mechanism
    advantages, split genes, 12
    progenotic genes, 85, 89
Urokinase, 14–19, 21, 46

Vaccinia virus, 14

Xenopus laevis, 119

Yeast
    mitochondrial introns, 127, 128
    DNA, 123
    triose phosphate isomerase, 107
    see also specific species
Y-shaped intron, 147